To Doris,

I'm so glad you were able to enjoy the Mara's beauty for yourself. I hope "The Marsh Lions" provides you with some happy memories of all you saw and enjoyed.

Very best wishes,

Jonathan P. Scott.

THE MARSH LIONS

I have just emerged from the depths of the great wide open spaces, from the life of prehistoric times, today just as it was a thousand years ago, from meeting with the great beasts of prey which enthrall one, which obsess one so that one feels that lions are all that one lives for. . .
KAREN BLIXEN *Letters from Africa*

THE MARSH LIONS

Text by

Brian Jackman

Photographs and drawings by

Jonathan Scott

Elm Tree Books London

To Brian's wife, Sarah Jackman
and Jonathan's mother, Margaret Scott

First published in Great Britain 1982
by Elm Tree Books/Hamish Hamilton Ltd
Garden House 57–59 Long Acre London WC2E 9JZ

Text copyright © 1982 by Brian Jackman
Illustrations copyright © 1982 by Jonathan Scott

Book design by Patrick Leeson

British Library Cataloguing in Publication Data

Jackman, Brian
  The Marsh Lions.
    1. Lions   2. Maasai Mara Reserve
    I. Title   II. Scott, Jonathan
    599.74'428   QL737.C23
    ISBN 0-241-10827-6

Typeset by Servis Filmsetting Ltd, Manchester
Printed in Italy by Arnoldo Mondadori Editore, Verona

Contents

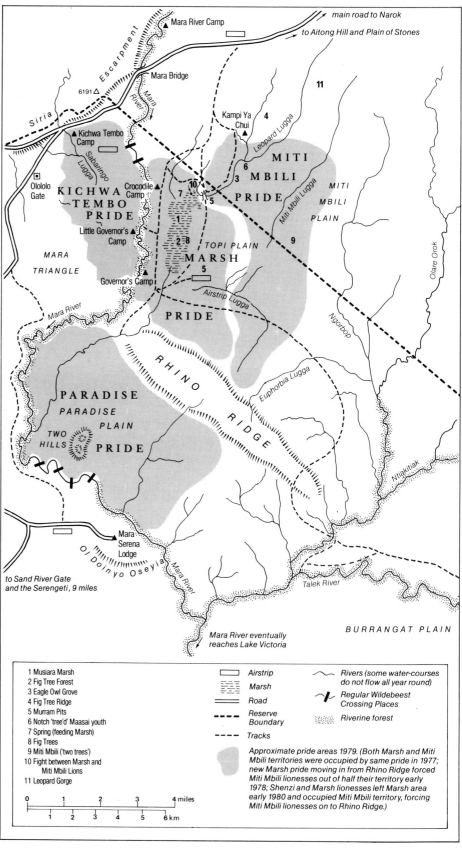

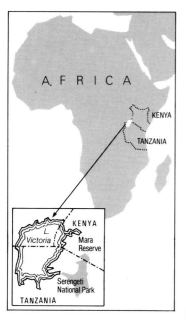

to Sand River Gate
and the Serengeti, 9 miles

main road to Narok
to Aitong Hill and Plain of Stones

Mara River eventually
reaches Lake Victoria

1 Musiara Marsh
2 Fig Tree Forest
3 Eagle Owl Grove
4 Fig Tree Ridge
5 Murram Pits
6 Notch 'tree'd' Maasai youth
7 Spring (feeding Marsh)
8 Fig Trees
9 Miti Mbili ('two trees')
10 Fight between Marsh and
   Miti Mbili Lions
11 Leopard Gorge

	Airstrip
	Marsh
	Road
	Reserve Boundary
	Tracks

Rivers (some water-courses do not flow all year round)

Regular Wildebeest Crossing Places

Riverine forest

Approximate pride areas 1979. (Both Marsh and Miti Mbili territories were occupied by same pride in 1977; new Marsh pride moving in from Rhino Ridge forced Miti Mbili lionesses out of half their territory early 1978; Shenzi and Marsh lionesses left Marsh area early 1980 and occupied Miti Mbili territory, forcing Miti Mbili lionesses on to Rhino Ridge.)

0 1 2 3 4 miles
1 2 3 4 5 6 km

An Introduction to the Maasai Mara Game Reserve

I N T H E hotel bars of Nairobi, and wherever the retired hunters and old
safari hands gather for a sundowner, you will often hear it said that the
Mara is the most beautiful of all the Kenyan wildlife strongholds.
Certainly it is the last place where huge concentrations of game can still be
seen in their old abundance. All the big five are here: elephant, black
rhino, buffalo, leopard and – above all – lion. Nowhere else are lions so
numerous. They roam the plains in prides up to twenty-five strong, and
the mature males, big-bodied and black-maned, are acknowledged as the
finest in Africa.

The lions share their wild kingdom with the wandering cheetahs and
hunting dogs, with jackals and hyaena clans. For most of the year prey is
plentiful. There are times when every horizon is a moving frieze of plains
game; giraffe, topi, kongoni, eland, impala, zebra and gazelle. And in
July the Mara and its surrounds become the scene of the greatest wildlife
spectacle on earth when thousands upon thousands of wildebeest and
zebra arrive from the Serengeti National Park in Tanzania, marching
north on their annual migration, and swelling the resident Loita herds
from the east of the reserve until the plains are black with animals.

It was the nomadic Maasai, those haughty warrior-herdsmen of the high
plains who measure their wealth in cattle and their courage in hunting the
lion with spears, who named this land between the Siria escarpment and
the Loita Hills. In *Maa* – the tongue which is a common bond between all
the Maasai clans – *Mara* means 'spotted'. So the Mara is the spotted
country, the stippled land, like a cheetah's skin; which is how the plains

The mature males, big-bodied and black-maned, are acknowledged as the finest in Africa . . .

must once have looked before it became the practice to fire the grass and burn off the acacia scrub and wild olive trees which dotted the open savanna.

It is the richness of the grasslands and the presence of permanent water which enables the Mara to sustain its vast populations of wild animals. Even in the harshest drought the Mara River never dries up. Born in the clouds on Mau Narok above the Great Rift Valley, the river begins its life as a series of rushing mountain torrents, carrying the rains that drench the cool cedar forests and bamboo jungles on a tortuous journey to Lake Victoria. By the time it reaches the plains it has become a lusty river, terrifying in full spate after the long rains of April.

For two hundred years the land belonged to the Maasai; but in 1891 disaster struck. Their herds were annihilated by rinderpest and the tribe itself stricken by smallpox. By the turn of the century the clans of the Mara Maasai were scattered. Dust devils twirled through their deserted manyattas and the country was left to the animals.

So it was that when the first Europeans arrived they found an earthly paradise inhabited only by teeming herds of game. The abundance of animals lured the trophy-hunters, and the next few decades saw the heyday of the professional hunters with their safari vehicles and tents and wealthy clients who were prepared to pay handsomely for the chance to bag a Mara lion.

After the Second World War the Mara lay wide open to exploitation. Shooting became an uncontrolled free-for-all. Biltong hunters cut huge swaths through the herds of plains herbivores, carrying off meat by the lorryload to Nairobi; and Wakuria tribesmen swept down from the escarpment to take skins and ivory. The poachers and honey-hunters fired the grass, destroying much of the ancient woodlands and, with the gradual return of the Maasai, cattle encroached as far as the tsetse fly would allow.

In 1948 a two-hundred-square-mile triangle of land between the escarpment, the Mara River and the Tanzanian border was declared a national reserve. Stricter laws to control the shooting of wild animals followed. In 1961 a further five hundred square miles was added and the whole area became the Maasai Mara National Game Reserve, a vast sanctuary in which Kenya's remaining plains game would be free to roam for all time. When the slaughter was stopped, and the reserve was finally established, fewer than ten male lions remained on the Keekorok plains.

Although human settlement is no longer permitted in the reserve, the Maasai are still in many ways as much a part of the ecology of the plains country as the wild animals. In the past they had always lived in comparative harmony with the game, despite the occasional tragedies in which Maasai died on the horns of a buffalo, or were chomped by a hippo or, more rarely, taken by a lion. Sometimes, in times of famine, the Maasai would kill buffalo and eland for meat, since they regarded them as wild cattle. But as they did not cultivate the plains, they felt no hostility to the zebra and antelopes, though their traditional tolerance has gradually been undermined both by the temptations of commercial poaching and the proliferation of the migrating wildebeest and zebra who compete for the same precious grass and water as their cattle.

Independence has brought big changes to Kenya; but deep in the bush the semi-nomadic life of the Maasai has hardly changed since the *Il-mee-oo-ibor*, the white strangers, first came to their land nearly a century ago. Cattle remain the backbone of the Maasai culture. Their economy still revolves around the brindled herds, gaunt but beautiful creatures with gentle eyes and swinging dewlaps; and their language contains no less than thirty words to describe the different subtle colours of a cow's hide.

For the Maasai, cattle are the supreme providers. They supply the sizzling flesh, spitted and roasted on iron skewers, which the warriors consume at ceremonial meat-feasts. They yield the milk which, mixed with blood drawn off from the animal's jugular, is a staple food. They give the dung with which the woman plaster the woven wicker huts, and provide the skins on which they sleep. Even their urine is used to wash calabashes when water is scarce. No ceremony can be performed which does not involve the use of cattle in some form or another; and, although the Maasai no longer go to war over a few stolen bullocks, and cattle-raiding is now officially forbidden, the possession of a fine herd still confers prestige on the elders and determines the marriage prospects of the young warriors in a society where the bride-price is always paid in cattle.

Life in the bush is as austere as their God, *Enkai*. But they believe He has chosen them among all the peoples to be the keepers of all the cattle of the earth. They are the Maasai: spearmen, lion-killers, drinkers of blood. So they cling stubbornly to their pastoral existence, roaming the plains in search of grass and water for their herds like the wild game whose freedom they share, and feel only pity for anyone unfortunate enough not to have been born a Maasai.

Maasai-land has changed, will continue to change, and the pace is accelerating. The great wheat ranches now reach out along the road from Narok almost to the frontiers of the Mara. Freshly-cultivated *shambas*, once confined to the green highlands of the Kikuyu country, spill down on to the arid moonscape of the Rift Valley floor where, only a few years ago, Maasai cattle browsed freely under the fever trees with giraffe and kongoni.

One day, perhaps, the thin proud nomads will disappear from the plains like the vanishing rhinos, which poaching has reduced in the Mara to a mere handful. But for the present, in the true Mara, from the Mara Bridge all the way down into the great emptiness of the Serengeti, only the ruts of old tyre tracks and the hum of a distant aircraft heading for Serena Lodge indicate that the twentieth century has ever come this far.

1

January 1978

Dark Mane's Pride

*. . . the unmistakable profile
of a roosting vulture . . .*

Bᴌᴀᴄᴋ shapes beneath the stars; boulders of rounded granite. The outspread canopy of an acacia tree; and perched among its branches, darker than the delicate tracery of leaf and thorn, the unmistakable profile of a roosting vulture.

Even in repose the hunched shoulders and fearsome hooked beak, designed for rending and disembowelling, gave the huge bird a malign and sinister appearance. On its breast, two livid medallions of bare flesh were exposed. These, and the white ruff from which the serpentine neck protruded, marked him as a Ruppell's griffon, one of the six tribes of airborne scavengers that sailed far and wide across the high plains of the Maasai country in an endless hunt for carrion.

With others of his kind, he often slept in the scattered trees that sprang from the rocks of the Siria escarpment, and the branches gleamed white with his droppings.

Below the roost the scarp fell away for a thousand feet in a tumbling chaos of boulders, thorns and strangler figs, a refuge frequented by leopard and baboon, hyrax and puff adder. The Maasai called this highest prow of the escarpment *Esoit Oloololo*: the Stony Place to Go Round. At its foot, hidden within a black ribbon of forest, the Mara River flowed between steep banks of crumbling clay, and across the river lay the deep grass oceans of the Maasai Mara Game Reserve, rolling south to the Serengeti.

Engulfed in shadow, the escarpment waited for daybreak. In the riverine woodlands the last shrill pipings of nocturnal tree-frogs faltered and died until hillside, river, forest and plains were held in a single silence. But as the first grey hint of dawn seeped over the eastern horizon a lion began to roar, and farther off, another, and yet another like an echo of the first, each drawn-out groan ending in a series of deep coughing grunts.

Halfway down the scarp, a prowling leopard announced his presence with a rhythmic, wood-sawing rasp. At once, a baboon barked an alarm, and the vulture awoke. No movement betrayed his awakening, only the reptilian glitter of a watchful eye. For some time the bird remained motionless, as if carved from the same volcanic rock that littered the surrounding slopes, while the sky grew brighter in the east and a tide of birdsong arose from the river.

Somewhere nearby a morning warbler began to sing. Dusky bulbuls chattered to each other, flaunting sulphur-yellow vents as they flitted among the rocks. Skulking unseen in the depths of a thorn thicket, a boubou shrike and its mate belled to each other in such perfect unison that call and response sounded like one bird.

Other voices joined in; the melodious fluting of black-headed forest orioles, the squeal of brown parrots flying over the treetops, the grating cries of guinea-fowl and francolin. And from all sides, over and over again above the mounting crescendo of chats and mousebirds, barbets, weavers, starlings and wood hoopoes, the joyful purring of ring-necked doves throbbed from every acacia top to greet the awakening African day.

Soon the sun appeared beyond the eastern hills, swelling as it rose until it became an enormous golden bubble that clung briefly to the blood-red horizon, quivering in the melting mist before breaking clear. The sunlight streamed across the plains, throwing into sharp relief the solitary wild

As the first grey hint of dawn seeped over the horizon, a lion began to roar . . .

. . . an enormous golden bubble that clung briefly to the blood-red horizon . . .

A few heavy wing beats were enough to loft him out above the river . . .

olive trees, the dark specks of grazing animals. The strange whistling yelps of the zebras carried far in the clear air as they thundered in panic through the shining grass, stampeded by marauding hyaenas.

As the sun rose, the shadows fled swiftly down the flanks of the escarpment. When the advancing sunlight touched the tree in which the Ruppell's griffon sat, the bird shuffled his dark wings and began to preen, tugging and probing among the broad overlapping shoulder coverts with the ivory hook of his beak. Despite his carrion-eating lifestyle the griffon kept himself fastidiously clean, and spent much time bathing in the shallow stretches of the Mara River, after which he would perch in a nearby tree with half-crooked wings hung out to dry.

Grooming completed, the big vulture sat in the tree, allowing the sun to thaw the chill from his body; then he launched himself into space. A few heavy wing beats were enough to loft him out above the river. He felt the warm air rising from the plains and rose with it in a long sweeping glide that took him back towards the escarpment. Here, he knew, there would be thermals and buoyant updraughts. By riding the torrents of air flung up by the escarpment wall, by harnessing the winds, he was able to conserve energy, to climb thousands of feet in a matter of seconds and remain aloft for hours on end while he scanned the plains for food.

On the ground he was grotesque and ugly. To the tourists who came riding out over the plains by Land-Rover and mini-bus to click their cameras at the lions and cheetahs, he was an object of loathing, a defiler of corpses who would arrive at a kill with bare neck outstretched, hopping and flapping through the grass in obscene haste to compete with his fellows for the choicest hanks of viscera. But flight transformed the ghoulish griffon into a creature of matchless soaring grace.

Now, as he felt the turbulent press of air under him, he canted over on outflung pinions and allowed himself to be carried upwards in a slow spiral until he was no more than a small black star in the sky. At around six thousand feet he found a high-altitude airstream and levelled out, flowing with it towards the sun. His eight-foot wingspan, exceeded only by the less numerous lappet-faced vultures, carried him effortlessly through space.

He was not alone in the sky. In every direction he could see the drifting silhouettes of other griffons and white-backed vultures. He heard the querulous keening of kites, the bark of a bateleur, and once, the rushing fall of a peregrine falcon as it stooped on some luckless dove. Tawny eagles and augur buzzards shared his remote vigil, driven by the same overpowering pangs of hunger.

Only the bateleurs, black and white eagles with coral-red legs and stumpy tails, seemed unmotivated by anything except the sheer joy of flight. Long ago the Frenchman Levaillant had watched them swaying and balancing on the thermals, wing-tips spread like extended hands, adjusting to each sudden shift and nuance of the wind with fingertip precision, and had called them *bateleurs* – a slang word for tightrope walkers.

But the griffon ignored the carefree bateleurs. He was looking for the tell-tale signs – trotting jackals, a pride of lions, a distant vortex of descending vultures – which might betray the presence of a kill. With his binocular vision, ten times more powerful than a man's, he could spot the slightest movement. Even at six thousand feet he could pick out a spring hare in the grass or a hyrax sunning itself on a rock.

Below him the golden plains rolled down into Tanzania and the acacia woodlands of the Serengeti National Park, beyond the Great Sand River. The migrating wildebeest which had remained in the Mara throughout the summer had crossed the Sand River three months ago. They had completed their long trek through the northern woodlands, running the gauntlet of the woodland prides, and were now back on the short grass plains around Lake Lagarja in the south-east corner of the Serengeti, where their calves would be born in February.

The departure of the huge Serengeti herds had left the plains strangely empty. The grass, thick and lustrous when the migration arrived in July, and so tall that a lion might walk through it undetected, was now reduced to stubble. The short rains of November brought a brief respite. Fresh grass sprang up where showers touched the earth, but the clouds had dispersed more quickly than usual, and drought gripped the land once more, causing grazers and predators alike to forsake the gasping plains, creating a general movement of game towards those water-courses and river margins which still held a faint flush of green.

Even the topi, who preferred the safety of the open grasslands, were driven to frequenting the winding luggas and forest edges where lions and leopards lay in ambush. Only the Grant's gazelles, whose extraordinary metabolism enabled them to go for months on end without drinking, seemed oblivious to the drought. Often they would not bother to seek whatever meagre shade existed on the plains, but would stand motionless in the noonday heat as if lost in some profound reverie.

And still there was no sign of a kill. With mounting hunger the griffon veered away from the lifeless grassland and drifted once more towards the river, where a wide green stain of marsh spread along the edges of the forest. It was now almost two hours since sunrise.

*   *   *

Earlier that morning, at about the same time as the griffon left his roost on the escarpment, a lion had stood on a stony ridge overlooking Musiara Marsh and roared at the dawn. He was a male in his sixth year, a magnificent specimen in prime condition, with the true dark mane of a Serengeti lion. Where his mane grew thickest it was a rich glossy black that spread from his forehead to between his shoulders. There was also much black in his chest hair and prominent black tufts on his elbows.

The black-maned lion was well-known to the safari camp drivers, who would usually find him somewhere in the vicinity of the Marsh, sprawled full-bellied in the shade of a wild fig tree, sleeping off the effects of the previous night's kill. Having lived all his life within the reserve, he had no

A wide green stain of marsh spread along the edges of the forest . . .

fear of the Land-Rovers that rattled within yards of where he lay, or of the tourists who craned through the open-roof hatches to take his picture. They were no more troublesome to him than the yellow hippoboscid flies that clustered on his stomach during the heat of the day.

The first sight of wild lions never failed to impress the tourists. They saw them moving through the blowing grass with that easy, loose-limbed swagger, or came upon them at rest, far out on the heat-stunned plains, sprawled under the flimsy canopy of a wild olive. Even in repose the sight of their slumbering forms produced an unconscious quickening of the pulse. From the safety of their vehicles the tourists saw how the muscles rippled under the sleek tawny hides. They noted the broad muzzles, the strength of the jaws, mouth half-parted to reveal the fearsome lower canines, the surprising pinkness of the tongue. They gazed at the massive heads of the males, shaggy maned and heavy with dignity, like the bearded profiles of archaic gods, and felt instinctively that they were in the presence of the ultimate carnivore.

The local safari camp drivers knew all the lion prides of the western plains. Months of constant observation broken only by the rains had made them thoroughly familiar with their fluid and complex family hierarchies and the territories they occupied. Where tourists merely saw lions, the most expert guides recognized each individual animal, as a shepherd knows his sheep. Several even had names, usually relating to some physical characteristic, which made it easier to identify individual animals when the subject of lions inevitably arose around the campfire at the end of each day's game-viewing. So the big black-maned lion of Musiara Marsh was known in the camps as Dark Mane.

The pride which Dark Mane had joined two years earlier consisted of seventeen lions who roamed a wide area extending deep into the open grasslands around Miti Mbili – the two venerable trees which grew side by side on the otherwise treeless expanse beyond Topi Plain, forming a distinctive landmark. Dark Mane shared the sovereignty of the Miti Mbili pride with two other adult males. One was a grizzled lion with a battle-scarred face and a mane as rich as wild honey that grew darker where it spilled across his shoulders. Smaller than Dark Mane, he was past his prime, and his age showed in his teeth. When he panted, slack-jawed in the heat, his open mouth revealed that he had lost some of his incisors and that his powerful canines were becoming blunt and yellow. He was known as Old Man.

The other male was a much younger animal with the sparse tousled mane of a four-year-old. He was lean and scruffy, and had a central tuft of pale hair that stood up on his forehead in a way that seemed to accentuate the impression of wildness and unpredictability. He was easy to identify by the cleft which disfigured his right nostril, the legacy of some quarrel over a kill long since forgotten. In spite of his youth he was already a huge beast – from nose to tail-tuft he measured well over nine feet – and he was known as *Mkubwa* – Swahili for 'big'.

They felt instinctively that they were in the presence of the ultimate carnivore . . .

. . . Old Man – a grizzled lion with a battle-scarred face and a mane as rich as wild honey . . .

. . . Miti Mbili – the two venerable trees which grew side by side . . .

. . . Mkubwa – an impression of wildness and unpredictability . . .

Tolerated by the males – perhaps because together they could intimidate him and thus did not regard him as a real threat to their authority – he was always something of an outsider, never fully accepted by the pride females. The lionesses feared his size and uncertain temper, and more than once his haunches had felt the red rake of their claws.

There were four mature females in the pride. Young Girl, the youngest, was a beautiful lioness with broad features, large eyes and thick, soft fur that was almost pure white under the throat.

Notch – so-called because of a distinctive nick in her right ear (the legacy of a quarrel with Mkubwa over a warthog kill) – was a dusky virago with a notoriously short temper. Unlike the majority of lions in the reserve she was always uneasy in the presence of safari vehicles. If they approached too close she would crouch in the grass with ears laid back and teeth bared, lashing her tail in anger. On one occasion she had charged a Land-Rover, biting and clawing at the bumper and terrifying the tourists. After that the drivers were always careful to give her a wide berth.

Notch's half-sister, Shadow, was exactly the opposite in temperament. She was as timid as Notch was aggressive, and always seemed to keep in the background, especially when Mkubwa was present. His first act on joining the pride had been to kill her two new-born cubs. He had found the small, strange lions with their unfamiliar smell and mewling cries, lying in a patch of long grass where their mother had hidden them. One he had killed on the spot with a single bite. The other he had chopped through the neck and shaken like a rat before walking away with the dead cub dangling from his jaws. He had carried it aimlessly for a few yards before dropping it, and then, after one sniff at the little furry body, he had slumped down in the shade and fallen asleep.

Notch was a dusky virago with
a notoriously short temper . . .

Such acts of infanticide, so cruel and senseless in human terms, are commonplace among lions. In the harsh and unsentimental world of the African bush there are sound practical reasons for this behaviour. Mkubwa had no interest in offspring sired by a male he had helped to banish from the pride. If Shadow's cubs had survived, she would not have given birth again until they were nearly two years old, whereas the trauma of losing her cubs would quickly bring her into oestrus once more: she would then be receptive to his advances. It is in the interest of all new males to beget as many of their own cubs as possible during their relatively brief reign as pride overlords. In these circumstances, cub-killing simply hastens the opportunity of fatherhood.

Therefore Mkubwa had not killed in anger. Nor had he killed out of hunger. He had only responded to an urge within himself, instinctively acting as his forebears had done, so that his own dynasty should stand a better chance of survival. Yet on this occasion, although Shadow had indeed come into oestrus again, she would not mate with Mkubwa, but had sought instead the company of Old Man, and the killing of the cubs only served to increase the wariness with which all the lionesses regarded the troublesome young male.

The fourth adult female of the group was the most striking in appearance. She was the big lioness called Old Girl, much scarred about the face and body, and instantly recognizable by her size and pale pelage, which had grown silvery-grey with age. Now in the second decade of her life, she had outlived all the other lionesses who were in the pride when she was born. Four times in her life she had seen the mastery of the pride change as successive groups of rival males had fought for possession of her pride's territory. Four times, at least, she had given birth, once under the same ancient acacia tree where she herself had been born at the edge of

Their strength is increased many times by their habit of consorting in prides . . .

the Marsh. Notch and Shadow were her daughters. Her last litter of three cubs – all females – were now nearly three years old. They were not yet sexually mature – their small nipples and slim white bellies showed they had not yet produced cubs of their own – but they were fast approaching the time when they would leave the pride to fend for themselves.

The remainder of the pride consisted of seven cubs belonging to Notch and Shadow, who had given birth almost simultaneously. There were four males and three females, but as they were so similar in age and appearance, and had been suckled by both mothers irrespective of parenthood, it was impossible to tell which cubs had been born to Notch and which to Shadow.

Lions are unique among the big cats. They are not only the most powerful of the plains predators. Their strength is increased many times by their habit of consorting in prides. Food is the factor that binds them together. Unlike the reclusive leopards who skulk in the thickets, seeking their prey by stealth and surprise; unlike the solitary and fleet-footed cheetahs who rely on their devastating speed to run down gazelles on the open plains, lions cannot match the agility of the spotted cats. Their very size and strength makes them appear heavy-footed in comparison; and this the lions overcome by living and hunting in social groups. They are masters of the ambush, most effective when two or three lionesses hunt together.

The pride system has another important advantage. Hunting in numbers enables lions to kill much larger animals than the bushbuck and impala which the leopards prey upon, or the Thomson's gazelles and wildebeest calves which fall to the cheetahs. Adult zebra, wildebeest and warthogs are the lions' favourite prey, but given the need and the opportunity the Miti Mbili pride would often pull down one of the full-grown buffalo which lived in large herds along the forest edges near Governor's Camp. Such a kill would provide meat for the pride for several days.

The size of each pride varies, and is determined by the abundance of prey, the suitability of the terrain and the presence of water. Of the half-dozen prides which lived within a morning's game drive of Governor's Camp, the smallest was the Leopard Gorge pride, to the north of the Marsh, with only seven lions, and the largest, the twenty-strong pride on Paradise Plain, to the south.

An average pride consists of perhaps four or five lionesses with cubs of varying ages, and two or three attendant males. Each family group lives within its own pride area, a well-defined territory whose boundaries the males regularly mark with vigorous paw-scrapes in the earth and sprays of urine tainted with scent from glands at the base of the tail, whose musky odour serves as an umistakable warning to intruding lions.

Although pride areas sometimes overlap in places, expanding or contracting in relation to the changing strength of the prides, they are usually surrounded by a kind of no man's land which is seldom seriously contested and which prevents the lions from wasting their energies in senseless internecine warfare. But within each pride area lies a jealously guarded familial core, a territorial heartland whose sanctity the resident lions will sometimes defend to the death.

The ancestry of any one pride is based on a matriarchal continuity. Its nucleus is always a group of related females – sisters, half-sisters, cousins, aunts – born and reared in the same pride. Many lionesses remain in the same pride all their lives, and spend all their years within the same pride area. In this way some prides acquire a history lasting several generations.

The adult males, too, are often closely related to each other, but generally unrelated to the pride females. Expelled as three-year-olds from the pride in which they are born, the sub-adult males form bachelor groups and live the shiftless life of nomads until they are old enough to fight for possession of a pride of their own. Once established, their period of tenure as pride-masters is relatively brief – perhaps only two or three years – before they in turn suffer the ignominy of being ousted by younger challengers.

The pride structure reveals lions as creatures with a surprisingly complex social order, but the pride itself is a loose-knit series of constantly changing relationships. Members of the pride do not consort with each other all the time, but often split up into smaller groups. Sometimes a lioness and her cubs will live apart from the main pride, and lionesses or

The remainder of the pride consisted of seven cubs . . .

Warthogs are among the lions' favourite prey . . .

He could hear other lions roaring . . .

sub-adults who have been close companions from birth regularly go off and hunt together. But they are always freely accepted by other members of the same pride, and seldom stray from their own pride area. Only the adult males, scenting the opportunity to mate with a receptive and unguarded female in an adjoining territory, may go adventuring beyond their own domain.

Such a situation had presented itself to Dark Mane during the latter half of 1977, when he met up with two unaccompanied lionesses in the tsetse-infested thornbush country which lay beyond the northern perimeter of his own pride area. He had at once mated with them both in turn, and each lioness gave birth to twin cubs, which they kept hidden in the rocks on Fig Tree Ridge, until they were all seen together with Dark Mane in January 1978. The cubs were then about eight weeks old.

Now he was returning to his old haunts, walking steadily southwards towards the Marsh. He ignored the impala who stopped grazing and watched him passing by, their burnished, golden-brown bodies quivering with pent-up nervous energy. A zebra stallion wheezed a warning, but did not flee. The herbivores knew the lion was not hunting.

Soon he had left the scattered acacias and come to the stony ridge that formed the northern boundary of his pride area. Below him the ridge sloped down to the edge of the Marsh. He saw the misty and familiar reed beds, the glitter of water in the red-gold light of dawn, and he roared at the plains beyond. He roared with eyes half closed and muzzle pushed forward, as if in deep concentration, forcing out each lungful of air with such power that tourists awakening in their tents at Governor's Camp, two miles away, heard his throaty vibrations and sat up in wonder. He waited until the echo had died away, and then his challenge crashed out again, the deep-drawn groans dying away in a succession of earth-shaking grunts. *Hii nchi ya nani? Hii nchi ya nani? Yango, yango, yango!* The Africans knew what the lion was saying. Whose land is this? Whose land is this? It is mine, mine, mine!

He stood on the ridge looking out on his world and felt the breeze lift the back of his mane in a hackled crest. Farther off, above the clamour of Egyptian geese in the Marsh, he could hear other lions roaring on Paradise Plain, and something more – the dull whine of a diesel engine.

Between the Marsh and the forest he could see buffaloes strung out like black beads across the wide bays of grass. There were waterbuck, too, with coarse shaggy coats and lofty knurled horns, and the ghostly grey shapes of elephants melting away through the trees. The diesel was louder now, but Dark Mane paid no attention. Vehicles held no fear for him. He gave no more than a cursory glance at the battered green hunting car that came lurching out of the thorn thickets and pulled up some thirty yards behind him. But when a man got out and moved towards him, he sank into a crouch, prepared to charge or flee. His own low rumbling growl of warning was the last sound he heard before the 300-grain Kynoch soft-nosed bullet slammed into him.

*   *   *

In the Mara all the carrion-eaters have their own ways of devouring the dead. Every species has its own particular niche which allows it to thrive alongside its fellows. The giant lappet-faced vultures can rip through hide with their meat-hook beaks. Sometimes they even hunt for themselves,

The giant lappet-faced vultures can rip through hide with their meat-hook beaks . . .

Vultures fight for leftover scraps with the darting jackals . . .

plummeting out of the sky to seize a gazelle fawn that had thought itself hidden in the grass. The griffons and white-backed vultures are adept at plunging their bare necks deep into a corpse to gorge on the entrails, while their smaller cousins the hooded vultures can pick through the carcass and fight for leftover scraps with the darting jackals. The bones themselves are carried off by the hyaenas. Such is the meticulous economy of nature that nothing is wasted. In time, all life on the African plains comes to this, ending in the bellies of vultures and hyaenas.

At last the patience of the griffon was rewarded. No sooner had he circled back towards the river than he spotted the first black specks of other vultures falling out of the sky towards the northern end of the Marsh. At once his keen eyes picked out the inert shape on the ground, and he began his own stupendous fall to earth, the wind hissing through his primary feathers as he vol-planed out of the blue.

When he was at treetop height he checked his speed, hung for a moment, long shanks dangling, talons spread for landing, then dropped into the grass. Even before he had clawed at the ground he was half running, half flapping towards the squabbling mass of vultures. Somewhere beneath them lay the decapitated body of Dark Mane.

2

February 1978

A New Alliance

THE DEATH of Dark Mane, illegally shot and decapitated by a trophy hunter, was a grievous blow to the Miti Mbili lions. The loss of their most formidable pride male had left them in a highly vulnerable position, and although it could not have been foreseen at the time, the disruption it caused would bring about dramatic and far-reaching changes among the neighbouring prides.

Even before the shooting, the security of the pride had been jeopardised by Dark Mane's amorous absences. With the mistrusted Mkubwa still considered as little more than a nuisance and a hanger-on, it was left to the veteran, Old Man, to repulse intruders. But the pride area was too large for a solitary male to hold for long. In the old days, he and Dark Mane had been inseparable companions. Together they had made the Miti Mbili pride a power to be reckoned with. Now that the partnership was broken there would be no shortage of potential challengers among the pride-seekers, bands of vociferous young nomads who would be quick to exploit such weaknesses and lay claim to the favours of the Miti Mbili females.

Throughout the month before Dark Mane's death, Old Man and the Miti Mbili lionesses – Old Girl, Young Girl, Notch and Shadow – had taken to hunting with their seven cubs in the vicinity of Leopard Lugga. The lugga was one of several meandering water-courses which threaded their way like veins through the grasslands. Half-choked with scrub, hidden between sunken banks deep enough to conceal a buffalo, signposted by a thin green line of trees, it came winding down from the higher ground which lay to the north and east, skirting the woodland groves of Kampi ya Chui and running through a culvert under a well-used game-viewing track before finally losing itself in the Marsh. During the rains it brimmed with floodwater, but the drought had long since sucked it dry, cracked its bed of black cotton clay and baked hard the pitted hoof-marks of buffaloes and antelopes around the few shrinking water-holes that remained.

The Leopard Lugga had been a favoured haunt of several generations of leopards. For several years now it had been regularly visited by a huge male with a cast in one eye. The Mara leopards had suffered badly at the hands of the poaching gangs who regularly infiltrated the reserve, but the wall-eyed leopard had remained at large. Somehow he had managed to evade all efforts to kill him, and with every passing year he became more wary and more cunning.

He was seldom seen. By day he slept in thickets and trees, or sought the cool caves of Leopard Gorge, a shallow ravine just outside the northern boundary of the reserve. His was a fleeting presence. For long periods he would disappear altogether, patrolling the outer reaches of his territory, until the gralloched corpse of a young impala dangling from one of his larder trees announced his return.

Sometimes, very early in the mornings, one of the grey Land-Rovers from the Mara River tented camp would inch its way down the grassy floor of the gorge. It moved so slowly that not even the rock hyraxes were aware of its approach, so that tourists scanning the rocks through binoculars might catch a glimpse of the leopard's dappled form as he lay indolently sunning himself, or see for the briefest second his fierce whiskered head framed in a thorn thicket's spiky shadows, or feel the glare of his mad yellow eye before he streaked for cover.

The wall-eyed leopard regarded all lions with a singular enmity. He had learned from long experience that lions meant trouble, and he avoided them because he recognized them as predators more powerful than himself. He avoided them because they were competitors who would chase him away from his favourite resting-places. He avoided them because of their rank lion smell. He avoided them because he knew he would have to surrender the familiar game trials where he liked to lie in ambush. But most of all he avoided them because he feared them, and sensed that they would kill him if they could. So when he came upon the Miti Mbili pride one morning, encamped replete and red-muzzled around a half-eaten buffalo under the trees of Kampi ya Chui, he choked back the growl that rumbled deep in his throat and slipped silently away up the lugga.

The pools in the lugga were green and stinking, but so long as the water remained the lions showed no inclination to move on. At night they would emerge from cover and fan out over the plains in search of prey, a menacing phalanx of ghostly grey shadows, padding through the starlight on large soft paws. They bushwhacked topi and they dragged warthogs squealing from their burrows, but they would always return before dawn, which usually found them resting beneath the solitary giant fig tree that grew just below Kampi ya Chui.

While the Miti Mbili pride were settled on Leopard Lugga, the entire southern flank of their territory lay unguarded. Here the plains rose gently towards a long hilltop cobbled with boulders and crowned with dense green stands of croton bush. Generations of black rhinos had browsed and tunnelled their way through the woody thickets, forming a maze of ancestral trails currently occupied by a female rhino and her calf.

On the other side of Rhino Ridge the red oat grasslands of Paradise Plain rolled down to meet small islands of stony ground, flanked to the east by thorny palisades of acacia, and to the south by a shallow stretch of the Mara River, where game trails left by the migrating wildebeest wound among archipelagos of inactive termite mounds, and sooty chats sang from the broken stumps of dead thorn bushes.

All this land between the ridge and the river belonged to the lions of Paradise Plain. The Paradise pride was led by an old male and his two comrades. This trio had originated in the Olare Orok area to the east of Miti Mbili, and had taken possession of the Paradise pride in a fierce running battle which had cost the old male a severe mauling. The pride had flourished. There were now six lionesses and eleven cubs, whose ages ranged from ten to fifteen months. With so many mouths to feed, they were active hunters, but they lived in a part of the reserve where there were always many buffalo, and had become skilled in the dangerous

. . . the two nomadic males known as Scar and Brando . . .

business of pulling down the old bulls who lived apart from the main herds.

Rhino Ridge itself formed a natural frontier for the feline clans. It defined the no-man's land which separated the Miti Mbili pride from the lions of Paradise Plain, and it was in this narrow corridor of neutral ground that the two nomadic males known as Scar and Brando had taken up temporary residence towards the end of 1977. Both were outcasts from the Paradise pride, driven out mainly by their own growing sense of restlessness, rather than by physical intimidation.

Their sojourn on the ridge attracted another ally – the misfit, Mkubwà. Unable to integrate fully with the Miti Mbili pride, he had also made several attempts to tag along with the Paradise lions. But the Paradise pride males would have no truck with Mkubwa, and when the three of them appeared together one morning, roaring as they trotted down the hillside towards him, his courage deserted him and he fled.

It was shortly after this incident that Mkubwa encountered Brando resting beside a zebra kill among the rocks on Rhino Ridge. At first the two lions eyed each other warily. Neither seemed prepared to give ground, but when Mkubwa lunged towards Brando he immediately rolled on one side with mouth agape in a gesture of submission. This was enough to satisfy Mkubwa, who turned and scraped the ground with his hind paws before lying down in the rocks close by. In their age and in many other respects the two young males were very much alike. Both animals were somewhat gaunt and rangy. Their tousled manes gave them the look of a pair of rakehells in search of mischief, and they shared in their natures the same wayward streak of devilment which had made Mkubwa so disliked

Brando immediately rolled on
one side with mouth agape in
a gesture of submission . . .

among the lionesses of the Miti Mbili pride. The solitary life of the nomad
was alien to their gregarious ways, and both lions felt a strange sense of
comfort in the presence of the other.

Together that night they were joined by Scar on the zebra carcass,
chasing away the hyaenas who tried to steal it from them, and although
Mkubwa moved off the next day by himself, a casual relationship had
been forged which would eventually bind all three together as the Marsh
pride males.

The two Paradise outcasts, Scar and Brando, were becoming increasingly fretful. Confined to the vicinity of Rhino Ridge, they were tired of
living beyond the pale. With every passing day, game became harder to
find. Yet on the plains below, deep in Miti Mbili territory, they could see
columns of zebra, topi and buffalo trooping into the Marsh to drink.

In the end, driven partly by hunger, but also by the rising tensions of
sexual maturity, they began to trespass along the southern fringes of the
Miti Mbili country. At first they were nervous. Their initial incursions
were in the nature of exploratory patrols, cautious probes into hostile
territory, carried out under cover of darkness. But when they sniffed at the
scrapes and pug-marks and boundary bushes and found no fresh scent
upon them, they became bolder and more confident, pushing deeper into
the forbidden grasslands of Dark Mane's kingdom until they no longer
bothered to withdraw during the day.

By the end of January they had established themselves at the very heart
of the pride area in the thickets of the Miti Mbili lugga and had still not
met with any resistance. When they sang to the dawn, their groaning
voices thundered over the plains and across the Marsh. The echoes of their

roaring rang down the luggas and reverberated along the walls of riverine forest, causing baboons to sit up and look anxiously about them, and bushbuck to bark in alarm. But when the air stilled, no challenge answered them, for Dark Mane was dead and Old Man was far away in the woods of Kampi ya Chui with a belly full of buffalo.

*   *   *

As February drew closer the drought increased its stranglehold. One by one the dwindling waterholes choked and died, until only the Marsh and the Mara River held much water. The hot winds that hissed like a death rattle in the barbed orchards of whistling thorn brought no respite. Everything about the landscape was scorched and shrivelled. In the riverine woodlands the leaves of the trees hung limp and listless. Cicadas gave the heat a voice, an interminable, mindless shrilling, strident and quivering like the air itself. Buffalo driven from their drying wallows took to standing in the river, or sought the shade in the deepest groves of the forest, and rain-birds cried *it-will-rain, it-will-rain*, as if to mock the thirsting land.

Two months of unremitting drought had reduced the grasslands to a dry and ghastly ash-blond stubble so brittle that it crumbled to powder beneath trampling hoofs and was sucked into the sky by twirling dust devils. Towards noon, the air turned liquid and ran like molten glass along horizons shaking with cruel mirages, in which Grant's gazelles far out in the shadowless plains appeared to be walking on lakes of blue water.

Often, during the breathless afternoons, fat cumulus clouds swarmed in the sky, piling up in dark towers of moisture that hooded the escarpment by sundown. The land cried out for rain, but no rain fell, and the sky cleared long before morning.

Now hunting was harder for all the predators. The wild dogs which had produced twelve pups the previous June had long since left their den on the Plain of Stones below Aitong Hill. They had not been seen since before the November grass rains, when they had followed the migrating herds as far as the Mara River. By now they were probably down on the Talek.

Every day on Topi Plain and in the meadow-like savannas bordering the Marsh, the old female cheetah known as Nefertari could be seen nosing through the grass in a constant search for the new-born fawns of Thomson's gazelles.

As for the wall-eyed leopard of Leopard Gorge, he was so hungry that he killed two silver-backed jackals and strung their remains in one of his larder trees, where their grinning masks dangled for several days until a tawny eagle found them.

For the lions, it was either feast or famine. After several days with empty bellies, the Paradise pride gorged themselves on a dead hippo which they found washed up on the banks of the Mara River. The Miti Mbili pride had eaten nothing but warthogs since the buffalo kill at Kampi ya Chui. Daylight found them sitting on an old termite mound in which a warthog had gone to ground. The lions took it in turn to try and dig the pig from the burrow into which it had fled, wheeling at the last moment and backing down the hole in order to face its pursuers with razor-sharp tusks.

Despite their bellicose behaviour the warthogs usually ran when lions appeared, but sometimes a boar would turn and face his tormentors, slashing and tusking a cat that failed to move swiftly enough, or was too slow in shifting its grip on the pig's hindquarters to a secure throat bite. The penalty could be a gaping facial wound of the kind which had disfigured Old Man in his early days.

The warthog's burrow had been excavated by an aardvark, the shambling ant-bear of the dry savannas, a shy nocturnal creature armed with powerful digging claws which it used to break into the termite citadels, burrowing deep inside the mounds to scoop up the insects with its long sticky tongue. The plains were riddled with their tunnelings, which provided a perfect refuge not only for warthogs but for hyaenas, jackals and a host of smaller creatures.

Sometimes a boar would turn and face his tormentors . . .

As one lion tired of digging, another took its place, paws working furiously to try and uncover the terrified pig. Sooner or later it would be dragged screaming from the hole, but sometimes the pig would make a last desperate attempt to escape. When this happened it seemed as if the mound had suddenly burst open in an explosion of brown earth and flying bodies as the fugitive shot out of the hole, causing the lions to leap back in surprise.

The cats would rush after their quarry as the pig ran for its life, with ramrod tail held defiantly aloft. The pig moved surprisingly fast on its stumpy legs. Sometimes it would get away. But if it was cornered by more than one lioness, or outpaced by the bounding lions in the long grass, it would soon disappear under a flurry of bodies as the pride closed in. Then the squeals of the victim would be drowned by the rumbling growls of the lions who lay pressed together, flank to flank, around the twitching carcass, snarling and squabbling over a kill which would do little more than take the edge off their hunger. Often they were so hungry they dared not release their jaws from the pig in case they lost their meal to another member of the pride.

The cats would rush after their quarry as the pig ran for its life . . .

. . . a kill which would do little more than take the edge off their hunger . . .

But no matter how many pigs they killed, there was never enough meat for all, and the Miti Mbili lions were forced to pursue more dangerous prey. One day they moved into the forest near Governor's Camp and surprised a bull giraffe while it was drinking from the river. Giraffes are seldom attacked by lions. Not only are they the most watchful of all the browsers; their deceptively gangling legs are quite capable of delivering a crunching forward kick that can send a lion sprawling with its ribs stove in.

On this occasion, luck was with the pride. The bull was at his most vulnerable, front legs splayed out, long neck lowered to reach the water, when the lions hit him. Old Girl's jaws closed on his windpipe as the combined weight of Notch and Shadow sent him crashing on his side in the mud. By the time the cubs had arrived, Old Girl had choked the life out of him, and the entire pride fed on his carcass for three days.

The two nomads, Scar and Brando, were also wandering farther from the Miti Mbili lugga in search of prey, and it was while returning from one of these forays that Brando encountered for the first time the fresh tracks of one of the Miti Mbili pride. He sniffed carefully at the broad pug-marks imprinted in the dust, and his muzzle wrinkled with interest at the odour of fresh urine. His nostrils told him the spoor was that of a lioness. Furthermore, there was an additional smell interwoven with her musky body scent, a fainter but more exciting astringence which his keen nostrils unravelled as that of a female who had just come into oestrus.

The lioness was Young Girl, who had remained in the vicinity of the giraffe kill after her companions had left, and was now returning alone towards the forest groves of Kampi ya Chui. Her trail skirted the edge of Topi Plain and led down towards Leopard Lugga, and Brando followed, nose close to the ground, moving after her at an unhurried pace.

Below Kampi ya Chui the lugga was joined by a smaller offshoot, and their meeting place was marked by a dark grove of wild olives, a linear woodland which hid the twisting water-course for perhaps a hundred yards. For years the wood had been occupied by a pair of Verreaux's eagle owls, biggest of all African owls. Their nest, an untidy stick pile commandeered from the pair of hammerkops which had built it, had sheltered generations of owlets; and it was here, two hours after sunrise, that Young Girl flopped down in the shade of the nest tree, among grey pellets of fur, feather, bone and beetle shards cast up by the owls after their nocturnal feasts.

The birds were asleep when Young Girl arrived, but their flat facial discs concealed ears so sharply attuned that they could locate and identify sounds as faint as the patter of a mongoose in the grass, or the dry papery rustle of the blue and orange agama lizards as they scuttled over fallen tree trunks. When the owls heard the slow footfalls of the lioness, they were instantly awake, staring down from the tree with their red-rimmed drunkards' eyes. An hour later, when Brando also appeared below the nest, the male owl uttered a mournful *hu-hu-hu* of alarm, and the pair of them sailed off on soft grey wings towards Kampi ya Chui.

OPPOSITE: *The giraffe was at its most vulnerable . . .*

At first Brando kept his distance. He lay down in the long dry grass at the edge of a small clearing where he could see Young Girl under the tree on the opposite side. For an hour, neither cat moved. They lay there, watching each other intently, while flies buzzed in the heat and the sunlight which pierced the dense canopy of leaves fell upon their tawny flanks in small spots and patches, rendering them almost invisible.

Suddenly Young Girl rose to her feet and strolled obliquely towards Brando, who growled softly but did not move away. Only when he sniffed beneath her tail did she pace swiftly across the clearing as if about to flee. Then, inexplicably, she stopped, rolled over and lay wriggling for a moment as if she was trying to rub an itch on her back, all the time watching Brando intently. Growing increasingly restless, she repeated her kittenish rolling several times, while Brando followed her, nuzzling and grimacing with pleasure at her scent and unsuccessfully trying to mount her.

For half an hour their courtship continued in this fashion, until Young Girl finally trotted towards Brando with her tail held high, turned round and presented herself to him. At once he covered her, his jaws agape on the nape of her neck. Their mating was fierce and furious. Young Girl rumbled open-mouthed, and when Brando announced his climax with a strange, harsh miaow, she turned on him with a grating roar and raised her paws as if to strike at his face, forcing him to dismount and back away.

But soon afterwards, when both lions had rolled on their backs and then had lain down beside each other, Young Girl enticed Brando into mounting her again.

By the end of the third day the pair had mated nearly two hundred times, resting close together between couplings which seldom lasted more than twenty seconds. Their carnality consumed them utterly, blunting their normally sharp senses to the extent that they seemed unaware of the life of the plains that continued to flow around them. They took no notice when the eagle owls returned at dusk, ghosting through the glade on moth-like wings. They seemed oblivious to the presence of an old buffalo bull that almost stumbled over them as he emerged from the lugga and lumbered off with a snort of alarm. Although they had not eaten since their courtship began, they ignored the presence of an impala harem that passed within yards of where they lay hidden in the grass. The intensity of their exertions had left them spent and exhausted. But gradually, as their procreative urges receded, they were replaced by the familiar pangs of hunger, and when the sunset flared and faded over the escarpment on the fourth day, they left the grove to the owls and set off together towards Topi Plain.

The night's hunting proved fruitless, but shortly before daybreak they managed to kill a young topi which Young Girl inadvertently drove into the jaws of her mate. Brando had slunk unseen along the dried-up bed of the Miti Mbili lugga and came charging out on top of the panic-stricken antelope before it could turn aside. Ravenously they tore open the beautiful buff belly, gnawed into the lustrous plum-coloured flanks until

Brando followed her, grimacing
with pleasure at her scent . . .

their own bellies were sleek and distended and they could eat no more. Then they crawled into a croton thicket and slept.

*   *   *

Old Man lay resting under a fig tree at the edge of the Marsh. During the night he had returned to the giraffe his pride had killed, only to find it surrounded by hyaenas. Sensing the hopelessness of the situation, he retreated and left the forest at daybreak, crossing the murram track that led to the Governor's Camp airstrip, and disappearing around the eastern fringes of the Marsh.

Had he been hungrier, had he been younger, had there been fewer hyaenas, he might have tried to chase them away from the dead giraffe. But he knew the ways of the dog-faced scavengers, knew the strength of their jaws, which could crunch through a zebra's femur as if it was matchwood. Despised as cowardly carrion-eaters, hyaenas are also ruthless predators who can butcher a wildebeest as efficiently as any pride of lions. What they lack in speed they make up for in stamina and determination. Where lions or cheetahs may give up after a short chase, the hyaenas will lope relentlessly after their prey until they wear it down. They are killers who live and hunt in close-knit clans. They are the hunch-backed gangsters of the high plains country, and Old Man respected them, perhaps because in their hideous shambling shapes as they whooped and fought among themselves over their stolen kills, he dimly sensed the shadows of his own death.

So the old lion had left the kill, walking slowly through the sunlight of the open plains until he came to the Marsh. Knob-billed Ducks and Egyptian geese rose up with anxious voices as he sniffed among the tall tasselled reeds, searching for a place to drink. Afterwards he left the Marsh and sought the shade of the fig tree, where he remained, eyes closed in half-sleep, far into the afternoon.

He lay alone. His pride had moved back into the woods of Kampi ya Chui, but this did not worry him. When he was hungry again he would hear them hunting. He would hear the sound of killing, and would seek them out to claim his rightful share as a pride male.

. . . ruthless predators who can butcher a wildebeest as efficiently as any pride of lions . . .

For some time now, Old Man had been aware of the presence of the nomads, Scar and Brando. Every night and again at dawn he heard them roaring, and answered them with his deeper challenge. Twice while

Knob-billed Ducks rose up with anxious voices . . .

Brando was drinking from a muddy pool . . .

trailing his hunting lionesses, he had come across the tracks of the trespassing males, but had not pursued them. Without the support and companionship of Dark Mane, he had become nervous and unsure of himself. Confrontation was now only a matter of time. When it came, he would have to fight for his pride, and maybe his life.

When at last he awoke, the fig tree was throwing long shadows across the grass. He rose and stretched, arching his back in a vain attempt to ease the dull ache of advancing age. He felt heavy with sleep. Old wounds nagged at his haunches, and he carried his shaggy head low, as if bowed down by the weight of years. But as he walked out into the plains he was still a majestic beast, and there was now a sense of purpose in his stride.

When Old Man strode over the horizon, Brando was drinking from a muddy pool a few hundred yards from the spot where he and Young Girl had killed the topi. The two males saw each other at the same moment. Young Girl, who lay off to one side in the thickets of the lugga, saw Brando's body stiffen and followed his gaze to the grassy rise where Old Man had paused for a moment, outlined against the sky as he stood sniffing the air.

The lioness remained crouching in the thickets, motionless except for an involuntary twitch of her tail, but Brando leapt to his feet as Old Man roared and came trotting towards him. The nomad was confused. Inexperience held him transfixed, unable to decide whether to flee or stand and fight. Too late he whirled and began to run; but no sooner had he turned than he felt Old Man's claws hook into his flank.

Over and over they whirled, their movements almost too swift to follow as they ripped and grappled and snarled, seeking the bite that would end the battle. Very quickly the older lion's experience began to tell. He feinted, rolled and bit savagely into Brando's shoulder. Brando felt the fangs sink deep. He felt the strength going from him, but managed to shake free and lashed out in desperation at Old Man. In his prime not so long ago the veteran might have dodged the blow, but now his reflexes were slower, and he caught the full force of Brando's claws on the left side of his face.

Young Girl, still hidden in the undergrowth, heard Old Man's roar of pain, saw the two fighting males break apart, with Brando streaming blood from deep wounds in his shoulder, and Old Man shaking a head made terrible by the gash which had all but gouged out an eye. Before Brando could regain his breath and return to the attack, Old Man's fighting spirit had broken. Maddened by the pain of his mutilated face, he turned and fled. His days as a pride male on Miti Mbili Plain were over. Now all he could think of was escape. Over Leopard Lugga he went, dripping blood as he ran. He skirted the northern end of the Marsh, and did not stop until he reached the Mara River.

3

February-May 1978

The Long Rains

. . . repulsing him with unmistakable hostility whenever he showed too much interest . . .

I T WAS now mid-February, and no sign of the drought breaking. Brando's wounds were healing well, though he was still troubled by the four deep punctures which Old Man's canines had made in his shoulder. He walked stiffly, and when he lay down he licked gently at the scars.

Within a few days of the fight, Young Girl had left him. Her brief time of sexual receptiveness at an end, she became peevish, repulsing him with unmistakable hostility whenever he showed too much interest, and eventually she drifted off by herself to rejoin her pride.

Meanwhile, her old Miti Mbili companions had become even more nervous. Although they were unaware of the blood-letting which had ended Old Man's reign, his absence haunted them, not as a conscious sense of loss, but as a nagging unease heightened by the sinister and persistent presence of the ambitious young males. Now that Dark Mane was dead and Old Man cast out, the Miti Mbili lionesses could no longer function as a true pride. By rights, by the immutable laws of nature, their territory now belonged to Brando, Scar and Mkubwa. But of the four lionesses, only Young Girl had submitted, and she had accepted only Brando. Old Girl, Notch and Shadow had all rejected the newcomers. Had they been in season they might have submitted as Young Girl had done. Instead they acted as if they were afraid of the strangers and, fearing for their cubs, shepherded them once more into the safety of the Kampi ya Chui forest.

Here, during the night, they killed another buffalo and fought over its carcass, snarling at each other with heightened irritability. The heat and tension of the past few weeks had made them snappish and irascible – none more so than the short-tempered, ragged-eared Notch. It was during this unsettled period in the history of the Miti Mbili lions that an incident took place which was to make her notorious throughout the safari camps and Maasai manyattas of the western Mara.

*   *   *

In the first golden hour after sunrise, when the sentinel olive trees and flat-roofed acacias flung their long shadows across the grasslands, a Maasai youth was walking over the dusty plains. While *kileken*, the morning star, was still bright in the sky, he left the dung-plastered hut in which he had làin with his *in-toyie*, and prayed in the forest as harsh-voiced turacos with cyclamen underwings glided among the treetops.

The hut was an inverted dome, like a swallow's nest. Inside it was dark and warm, steeped in the mingled smells of milk and woodsmoke, and of the ochre and mutton fat with which the youth and the girl had anointed their bodies. She had not wanted him to leave, but he had an important journey to make. Besides, was he not a moran, an *ol-barnoti*, or warrior-apprentice? Had he not unflinchingly borne the painful circumcision, the traditional rite of passage which had ended his boyhood, not crying out when the painted 'Dorobo', a cattleless wanderer from the Nguruman forest, had pared away his foreskin with a honed iron knife?

Now, as he strode across the plains, arrogance shone from him. He moved with a casual, coltish grace that was almost effeminate, as was the way in which he sometimes tossed his head, shaking the long locks which had been ochred and greased with sheep fat, teased and braided into an elaborate coiffure which swung freely down his back. But his thin brown body belied the legendary toughness and stoicism of his tribe.

Their territory now belonged to Brando, Scar and Mkubwa . . .

In his right hand he carried a long, sword-bladed spear, its shaft of pale wood balanced on his shoulder. The lobes of his ears, which had been pierced when he was eight years old and plugged with wads of leaves to enlarge the holes, were adorned with rings of beads and bone. Around his throat he wore a band made from the stomach-lining of a goat, from which hung a miniature pouch containing the fragrant root of the seyia reed. The smell of the reed pleased him, as did the mint-scented leaves of the *osinoni* bush which he and his age-mates used as an under-arm deodorant.

As he walked, the chill morning wind lifted the short cloak, the *ol-kila*,

He saw a vulture plummeting from the sky . . .

made from red cloth printed in Holland, which he wore about his shoulders like a toga, and with his free hand he drew it more tightly around his naked buttocks. The spear, the red toga and the coppery helmet of hair gave him the look of an ancient Roman warrior. Indeed, there were those who believed that the Maasai were descended from Arabian mercenaries recruited to fight in Egypt, who had fled south after the debacle of Anthony and Cleopatra; a lost legion taking with them their Roman ways. The only discordant items were his sandals, cut from the tread of a cast-off motor-car tyre and bought in the *dukas* of Narok.

For the fledgeling warrior, this was perhaps the most carefree period of his life on the high plains. Now that the first year of his warriorhood had passed he was at liberty to spend the nights with his girl-friends, or carousing with his age-mates at one of the month-long *olpul* or meat-feasts staged at specially constructed camps far out in the bush. The Maasai believed these long bouts of meat-eating gave the warriors strength, and it was to join such a feast that the youngster was now hurrying as he cut across the northern corner of the reserve.

In his mind's eye the moran could already see the camp, built in a clearing on the banks of a tributary of the Talek River. There, a chosen bullock would have been led to slaughter and fires prepared to roast the meat which he and his companions would eat at the end of the day, gorging until their bellies were swollen and rounded.

So strong was the image of roasting meat – he had taken nothing that morning but a calabash of smoke-flavoured milk – that he could almost smell the spluttering fat, the seared flesh. But a dark shadow drifting over the yellow grass in front of him brought him out of his reverie. He looked up and saw a vulture plummeting from the sky to join a growing spiral of black specks, denoting the fresh discovery of a kill on Topi Plain.

The Maasai call vultures 'the birds of the warriors', recalling the past, when cattle raids and blood feuds left a toll of speared corpses on the ground, and the big black birds would gather to peck at the soft eyes of the dead.

The arrival of the vultures caused him to become instantly alert. He gripped his spear tighter and looked around him for signs of dangerous game. Skirting the forest of Kampi ya Chui, he peered into the green gloom beneath the trees. Here, he knew, lurked *ol-kinyalosho*, the 'calf-eater', as the leopard was known to the Maasai, and *ol-arro*, the buffalo, 'he whose horns point downwards'. But all he saw was a herd of zebra stampeding away across Leopard Lugga. Nowhere could he see the twitch of a dark ear in the grass which might betray the presence of a crouching lion, and after a brief pause he continued on his way.

The Maasai word for lion is *olng'atuni*, but it is never uttered aloud lest the mention of its name should tempt the fates and invoke a visit. So the lion is always spoken of obliquely, as *olowuaru-kitok*: the big carnivore.

It is not dread but respect that makes the Maasai set the tawny cats apart from other dangerous animals. After centuries of living among lions, they know only too well the speed of the charge, the power of the jaws which can break a man's limb as if it were a rotten branch, and the cunning of the man-eater who comes in the night, or sometimes carries off herd-boys in broad daylight. That is why the lion is regarded as the most formidable adversary of the plains, why the *Olamayio*, the ritual lion hunt, is the greatest test of bravery a Maasai warrior can face.

Now that Europeans come in their hundreds to see the big game, the Maasai are forbidden to hunt lions, but old habits die hard, and every once in a while, perhaps out on the Loita Plains, the word goes round, the moran gather with their spears and painted buffalo-hide shields to chew the leaves of a certain small plant until their blood is hot, and set out in pursuit of a black-maned lion.

When they find it they encircle their quarry, chanting and taunting and yelling abuse. No one hangs back. As the hedgehog of spears closes in, every moran wants to be the one to receive the charge of the *olowuaru-kitok* as he tries to break out of the circle. When it happens – the chilling roar, the tawny blur – the warrior goes down under his shield, trying to protect himself from claws which can tear a man's face from his body, while the rest of his comrades rush forward, stabbing, stabbing.

Afterwards, the warrior singled out for the charge is honoured by the most prized of all trophies. Henceforward he is entitled to wear the lion's dark mane as a ceremonial head-dress. Then the hunters return home in triumph, and there is dancing; the men with finger-painted bodies, the girls with their beads and bouncing breasts, facing each other in stamping lines, chins up, heads thrust back as if hypnotised by the pumping pelvic rhythm and the chanting of the moran, whose deep asthmatic groans seem to echo the voice of the vanquished lion.

This much the young warrior-recruit remembered from his own childhood when a lion had been speared not far from the spot where the tented camp now stood by the Mara River; and the thrills of the hunt had been recounted many times by the elders until he, too, longed to prove his manhood by blooding his spear on a Mara lion.

By now he had reached Leopard Lugga and was just clambering up the bank when he heard the first warning rumble. So engrossed was he in his daydreams of glory that at first he felt he must have imagined it, but there was no mistaking the second growl. Even as he stood there, unable to control the lizards of fear that scuttled through his stomach, a full-grown lioness emerged from a croton thicket not more than a hundred paces away.

Notch was in an ugly mood. Separated from the Miti Mbili pride during an unsuccessful hunt, she had run into the unwelcome presence of Mkubwa and been forced to make a wide detour over Topi Plain, until she had found a sheltered place in which to lie up through the day.

It was sheer misfortune that the boy had chosen to cross the lugga at the very spot where she was hidden. Normally she might have leapt to her feet with a startled *whoof* and fled along the lugga; or remained concealed, silently watching from behind her screen of dappled leaves until he had disappeared from view. But on this occasion she felt threatened by the sudden intrusion of a human on foot. Fright dilated her eyes, filling them with baleful light. The upper lip curled back to reveal yellow canines as thick as thumbs. She growled again, a low guttural rattle that seemed to come from deep inside her chest.

Slowly she sank to the ground, gathering her hind legs beneath her. Seconds passed. Her whole body tensed as she prepared to launch herself across the intervening space, but still she did not charge. The restless twitch of her black tail tip betrayed her indecision. What held her back was the man-scent. It hung in her nostrils, awakening ancestral fears that not even her anger could overcome.

For a moment they faced each other as if transfixed. In the silence the insistent cooing of ring-necked doves carried softly down the lugga. Then, as if in a dream, she was coming for him. He shook his spear and screamed at her. She stopped, having covered half the distance that separated them, bristling with fury.

There was only one possible refuge. A short distance from the lugga stood a stunted acacia, little more than a bush. Its branches were a mass of vicious finger-length spines, but the moran forced his way into it and drew his legs up beneath him. This was the extraordinary scene the tourists from the Mara River tented camp encountered when their Land-Rover crossed the lugga a few moments later. Had they not arrived it is almost certain that the Maasai would have been killed or at least badly mauled. The bush itself was no more than ten feet high, and it could have been only a matter of time before the lioness dragged him down. Even when the Land-Rover approached, Notch was reluctant to leave. Only when the vehicle drove straight at her did she give way and disappear into the lugga.

When she had gone, it took a long time to coax the moran down from his tree and to extricate him from the thorns. For a while he just clung there, blood streaming from his trembling body where the sharp spines had impaled him.

Afterwards, Notch gained a new name at Mara River Camp. From that day on she became *Mama Kali*, the Angry Woman.

*   *   *

As the time of the long rains drew nearer, the weather became sultry and oppressive. The highland air lost its clarity. Distant hills which normally stood up sharp and blue across the plains were now obscured by haze. A fine suspension of dust hung in the sky, and palls of smoke billowed up over the Siria escarpment where the Maasai had fired the grass, creating brief but apocalyptic sunsets.

Far to the south, in the lowest reaches of the Serengeti, the rains had already broken around Lake Lagarja, where the wildebeest were gathered on the short-grass plains, timing their season of birth to coincide with the first rich flush of green growth. But it might be several more weeks before the slow march of the massed rain-clouds reached the Mara.

Everywhere, out on the dun-coloured plains and along the river

margins, the air was heavy with the sweet scent of acacias. Fierce black bees swarmed out of the woodlands to gather nectar from the creamy blossoms, while honey badgers plundered their wild hives, and bee-eaters gorged themselves among the branches. Time after time they dived from their treetop perches, wings fluttering like opaque bronze fins in the sunlight as they picked off the bees in mid-air, and returned to beat their victims senseless against a branch.

Mkubwa, lolling in the shade of the flowering acacias, no longer bothered to look up when he heard the bee-eaters. From dawn to dusk their shrill cries and the dynamo hum of the bees filled the lazy vacuum of his days. He lay outstretched with eyes closed, white belly upwards, paws curled, like a huge fireside tabby. Yet even in his drowsiest moments there was a part of him which never slept, and his round ears twitched constantly as he monitored the noises which came to him from the surrounding plains.

Above the sound of his own panting he heard the monotonous descant of wood doves, the bark of a zebra stallion defending his mares against a rival, the sepulchral booming of ground hornbills, big black birds with scarlet wattles, strutting like turkeys across the grasslands, to snap up grasshoppers which they tossed in the air with their pickaxe beaks and swallowed.

Palls of smoke billowed up over the Siria escarpment . . .

Bee-eaters gorged themselves among the branches . . .

. . . ground hornbills, big black birds with scarlet wattles . . .

Presently a new sound came to him: a curious, repetitive click. Although the young lion had heard it before, it was unusual enough to make him roll over and open his eyes. Slowly he raised his head and peered through the nodding grass-heads.

Out of the heat haze a large animal was approaching. Humped and dewlapped like a Maasai ox, its heavy bovine profile was crowned by spiral horns, swept back in line with its tapering muzzle. Its soft dove-grey flanks were elegantly marked with white pin stripes, and a red-billed oxpecker rode on its neck, searching for ticks in the folds of its hide.

The bull eland had been travelling steadily since dawn. Now, as he stopped to stare, the clicking noise stopped too. The sound was made by the tendons in his forelegs. At every step they gave out a sharp, twig-snapping crack that could be heard up to half a mile away, giving warning of approach to other bulls.

Eland are the biggest of the antelope tribes – a bull may weigh close to a ton and stand six feet tall at the shoulder. Too big to hide from predators, they rely on their exceptionally keen senses. Of all plains game, they are the shyest. When they browse they move upwind, carefully sniffing the air for danger, their far-seeing eyes constantly on the alert. So wary are they that a safari vehicle can seldom approach within four hundred yards before they are off, running like the wind for mile after mile.

They are true nomads, wide-ranging wanderers of the empty plains and high rolling ridges. In the dry season they go for weeks without water, obtaining sufficient moisture by browsing at night on dew-laden bushes, and so perfectly have they evolved in their sun-scorched world that they even possess a mysterious ability to stop themselves sweating, conserving still further their precious body fluids. Nevertheless, when drought manacles the grasslands, the eland migrate towards the gallery woodlands and riverbanks where more nourishing browse sustains them until the rains.

That was where the lone eland was now heading. For several years he had consorted with a small troop of other males that were occasionally seen to the east of Paradise Plain, roaming a territory so vast that there was no point in defending it. But with advancing age he had become more solitary, and no longer sought the companionship of the herd.

The old bull was suspicious. After twenty seasons on the plains, he had learned to be wary of shady trees which lions might favour in the heat of the day. He, too, had been seeking a shady place in which to pass the noon, but something about this tree made him uneasy. A sudden movement, perhaps the flick of a dark ear, had stopped him in his tracks when he was still some way off. For a long time he stood and stared, occasionally scything his horns through the withered grass, but Mkubwa was well hidden and the wind did not betray his presence. Only when a crowned plover began to scream and dive over the lion's hiding place did the eland take fright and canter off towards the upper reaches of Musiara Marsh.

Mkubwa watched him go, then rolled over on his back and closed his

A bull eland may weigh close to a ton and stand six feet tall at the shoulder . . .

eyes. The eland's sweet cattle-smell awoke pangs of hunger, but the bull was now too far away for him to tackle.

Later, however, after the last sand grouse had come flighting in to the Marsh, falling through the twilight with unearthly throaty chucklings, Mkubwa found Scar and Brando near the spring. The eland had failed to spot the three dark shapes concealed in the grass as he made his way down to the water. When the lions charged forwards the bull was trapped. In vain he tried to flee through the muddy reed-beds. There was no escape. The huge antelope was killed and dragged to dry ground by the hungry males.

All night long the Marsh echoed to their growling and the crunching of bone. Towards dawn, after the lions had eaten their fill and dragged themselves off to sleep in a thicket, the hyaenas came whooping and squealing from Leopard Lugga to fight over the remains; and in the morning, the marabous descended from their treetop lookouts to snap up what was left.

The marabous are meat-eating storks which have acquired the carrion habits of vultures. Hideously ugly, with scabby heads, pendulous neck pouches and leering Skid Row eyes, they would stand for hours in the trees, watching for corpses in the river or surveying the Marsh for leftover kills. Between the river and the Marsh, the marabous seldom go hungry, and their sentinel shapes hunched in the trees near Governor's Camp are a permanent feature of the Maasai plains.

It was not often that fortune favoured the lions with a kill as big as an eland. Already they were learning that life as a trio brought more success than roaming alone. They were finding, too, that together they were able to take kills from the hyaena clans simply by charging headlong towards them with angry growls. So brazen did their banditry become that they lived for a while almost entirely by scavenging.

During their nightly wanderings they came less often upon the nervous fleeting shapes of the Miti Mbili lionesses. Increasingly Old Girl and her companions had taken to hunting farther out on the plains, resting by day on the rocky ridges where they could feel the breeze as they stared across the sunlit gulfs below. Their impassive gaze took in everything: the endless grass, the lonely trees, the ambling giraffes and scattered herds of game. Slowly, as the weeks passed, they settled there, feeling a growing sense of security now they were no longer harrassed by Mkubwa and his allies. The old order was changing. As if by unspoken consent, the territory once ruled by Dark Mane and Old Man was being divided. With Brando, Scar and Mkubwa now undisputed masters of the Marsh, the Miti Mbili lionesses grudgingly retreated into the eastern half of their former haunts, consolidating on the Miti Mbili Plain itself, where the two trees grew together, and where the Marsh males seemed strangely reluctant to follow. For a while – though not for long – the situation was extraordinary: the Miti Mbili lionesses had no pride males, and the Marsh males had no females.

Towards the end of the wildebeest migration in 1975, Old Girl had

given birth to four cubs under a giant hollow acacia at the eastern edge of the Marsh. One cub was almost immediately lost, presumably killed by hyaenas, but the other three, all females, survived. In appearance they were remarkably similar, lithe and big-boned with pale, unscarred coats. When they became sub-adults they had left Old Girl to follow a nomadic existence, for the pride already had too many lionesses. Mostly they ranged along the slopes of Rhino Ridge, occasionally roving out on to the eastern plains where lived the black-maned Talek prides. During this time they became reunited with two other lionesses, known as the Talek twins, older cousins who had left the group a year earlier but had never entirely severed their links with the pride.

Together the five outcasts found greater security. They could no longer be driven off so easily by members of resident prides whose territories they sometimes crossed, and as their collective hunting technique improved, so did they feed more often and grow stronger.

Soon their presence was known to all the lions of the Talek, Paradise and Miti Mbili prides – and also to Mkubwa, Scar and Brando. Indeed, Scar had encountered one of the Talek twins in oestrus, with the result that she was now heavily pregnant and would soon give birth.

By February, the three Marsh sisters and the Talek twins were seen together for the first time on the banks of the Miti Mbili lugga. Their days

. . . the marabous, hideously ugly, with scabby heads, pendulous neck pouches and leering Skid Row eyes . . .

of exile were over. Unchallenged, they remained in the area, hunting between the lugga and the Marsh and, after Old Girl's withdrawal, behaved increasingly as if it was their own.

*   *   *

In late March the pregnant Talek twin produced her litter. There were only two cubs, and they were born under the stag-headed acacia tree which had been Old Girl's birthplace years earlier.

Soon afterwards the air cleared, the haze dispersed and a dry wind came rushing over the oat grass prairies. The wind ran along the ground like a great sigh, as if the earth itself could feel that the drought was almost over. Where the whistling thorns grew, dwarfed and stunted like unkempt midwinter orchards, the numberless dried and perforated galls in which fierce stinging ants had made their homes now rang in the silence of the day, a melancholy sound like the mindless sibilence of tiny flutes.

The scattered olive trees, clipped into curious parasol silhouettes by browsing giraffes, stood out darkly, like English holly, and the hills were no longer remote and enigmatic. Instead, as if by some trick of the light, they seemed to draw closer, resounding in their plum-coloured purity by evening.

Now the sun going north after the equinox drew the monsoon after it. The great anvil-headed cloud castles crept steadily closer, toppling into the blue while lightning forked and flickered along the southern horizon. The lions seemed to sense that the rains were coming. They roared in the iron-grey dawns and heard their voices answered by the sound of distant thunder. In the late afternoons they lay sphinx-like on the rocky ridges, manes streaming in the wind, gazing with gold eyes at the approaching storm.

Towards evening, elephants emerged in the muted light to move solemnly across the expectant plains. The sound of thunder unsettled the grazing game herds, causing the ever-skittish troops of zebra to go scudding across the grass at a mad gallop, keeling this way and that, like yachts in a stiff breeze, their sunlit stripes standing out brilliantly against the darkening sky.

When at last the drought finally broke, the first spots of rain as big as Kenya shillings, there rose the glorious smell of earth freshly slaked. Crinum lilies unfurled their pale pink trumpets amongst the tall grasses, and on Topi Plain a pair of crowned cranes faced each other with fluttering wings, rising and falling through the elegant ritual of their courtship dance. Everywhere, new life was returning. In the riverine meadows to the west of the Marsh, Nefertari the cheetah watched her three tiny cubs playing on the silver deadwood of a fallen acacia, oblivious to the hammering curtain which now hid the escarpment.

The rains did not fall continuously, but came in a series of showers and cataclysmic downpours interspersed with days of drizzle and sudden shafts of dazzling sunshine which made the earth steam warmly. The Mara River rose with frightening swiftness. Its brown waters burst their

banks, carried away trees, and drowned two Maasai who tried to cross it by holding on to the tails of their floundering cattle.

Elephants emerged in the muted light . . .

Flash floods roared down the luggas, forcing warthogs and dik-diks to flee before them, and sweeping Nefertari's three cubs to their deaths. To the east, the Talek River, which had been reduced to a chain of torpid pools, now raced in full spate through the gallery woodlands.

Almost overnight the Marsh became a lake, its neighbouring groves and woodlands islanded by the rising waters, forcing the Talek mother to carry her new-born cubs to higher ground. Everywhere the black cotton lowlands became a quagmire. Buffalo wallows brimmed to overflowing. The roads became impassable even to vehicles with four-wheel drive, and few tourists came.

The gift of rain renewed the plains. A mist of green now spread to the horizon, covering what had been nothing but dust and stubble. The savanna grasses had been waiting for this moment of cyclic rebirth. New shoots now sprang from the ground, growing at the rate of an inch or more

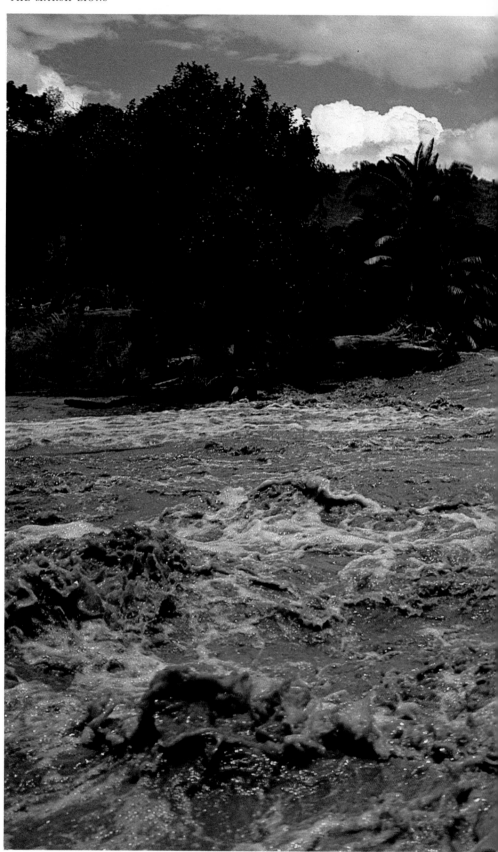

The Mara River rose with frightening swiftness . . .

every twenty-four hours, until the plains were waist deep and the endless meadows of the Mara rippled in the wind like young barley.

The rain and the grass: these were the fundamental elements on which the plains depended. The rain fed the grass and determined its growth, which regulated the numbers of herbivores the land could support. They in turn, according to their abundance, dictated the number of predators. So it had been since the Pleistocene; a precarious paradise, beautiful to behold, both in its economy and complexity, stabilised by the extraordinary diversity of interacting life forms that maintained its continuity.

Through the mysterious alchemy of photosynthesis, the grass transmuted the energy of the sun into sugars. The grass was eaten by gazelles and zebra and antelopes. Its sweet cud fuelled their movements across the plains. In return, their dung replenished the soil, while their flesh nourished the lions and leopards, cheetahs, hyaenas and hunting dogs. In time, when the predators themselves died, they, too, returned into the earth and re-entered the life-giving grass.

In the wake of the rains, when storm clouds trailed their dark veils beyond the Mau escarpment, the Mara was a glittering emerald shot through with the gleam of water that lay everywhere, in pool, marsh and lugga. Cycnium flowers sprang up as if from nowhere, stippling the plains with delicate white blossoms. The spicy scent of shrubs and marsh plants prickled in the nostrils, and over everything hung the rich meadow smells of warm grass and wet earth.

With the greening of the plains, the herds which had been concentrated along the forest margins of the Mara River now dispersed. Gaunt ribs swiftly disappeared under sleek and healthy coats as the game grew fat. Topi, kongoni and Thomson's gazelles nibbled eagerly at the tender new leaves of the star grass and guinea grass. Giraffes craned into the tall acacias. With leathery tongues as long as a man's forearm, they grasped the emerging foliage and plucked it from the spiny branches. Elephants, which had remained concealed for much of the dry season in the thick riverine forests around Governor's Camp, now emerged in broad daylight and marched across the plains, growling and trumpeting as they followed their leader, the wise old one-tusked Musiara matriarch. Sometimes in their journeyings they encountered the rhino calf which had lived on Rhino Ridge. During the rains its mother had been killed by poachers and the precious horn hacked out, to be smuggled by dhow to the Yemen. There it would change hands for thousands of dollars, to be carved into dagger handles for Yemeni sheikhs. Little of this wealth ever reached the poachers themselves. The real killing is made by the middle-men, the big-shilling bwanas in Nairobi and the Mombasa go-downs who have secured huge personal fortunes by robbing their country of its national heritage. So great is their greed that, in the space of a few brief years, the Mara's once prolific rhino population has been reduced to little more than a dozen animals. After her mother had been shot, the calf had left the ridge where the corpse lay rotting in the thickets, to wander away through the verdant grasslands, a forlorn and lonely figure in the midst of so much teeming life.

Throughout the Mara the rains unleashed a frenzy of activity among the birds. Red-necked with lust, hot-blooded ostrich cocks pursued their hens in absurd high-stepping dances. Kori bustards preened and strutted before their mates with swollen throats and uplifted tails. Buttercup clouds of weaver birds chattered and built their dangling raffia nests in riverside acacias. Larks and yellow-throated longclaws nested in the drier tussocks of the high plains, filling the air with plaintive cries. Crowned plovers laid freckled eggs in shallow scrapes on the bare earth and rose screaming around the heads of trespassing topi. *My love, forget-me-not*, crooned the red-eyed doves in the strangler figs, and fish eagles yelped as they sailed over the drowning Marsh.

The termites, too, were aroused. Deep in their subterranean citadels, they heard the tattoo of drumming rain and made their way towards the surface. There, after the workers had made special openings, the mounds erupted like geysers as thousands of insects poured into the evening sky. These were the alates, the winged ones, a race of future kings and queens who would never again taste the freedom of the air after this one brief nuptial flight. Once they had drifted back to earth, they would discard their wings and search about in desperation for a mate. Then, having paired off, they would return to their underworld to begin new dynasties and, in time, to raise fresh mounds across the plains. Only a very few of them would succeed. For the majority, the flight would lead only to waiting mouths: spiders and mantids; toads and monitor lizards; pearl-spotted owlets and bat-eared foxes. Swifts and nightjars caught them on the wing in their whiskered gapes; secretary birds sought them on foot in the grass, and baboons scooped them up in handfuls from the ground.

If the Mara could be said to have a springtime, it was now, in the aftermath of the long rains, when the grasslands were a lustrous green, when the whole country seemed to shine with a kind of new-minted freshness, and every grove, plain, lugga, marsh and hollow buzzed, seethed and pulsed with life.

Widow-birds in full breeding plumage fluttered over the swelling grass-heads, almost dragged down by the weight of their absurdly long black tail feathers. Swallowtail butterflies flipped through the forest glades and settled with a flicker of iridescent green and black velvet wings at the edges of muddy puddles.

Everywhere, the old law prevailed: eat and be eaten. Troops of banded mongooses ran nose-to-tail among the anthills, plunging like porpoises through the glistening grass, taking whatever they could find – scorpions, millipedes, snakes, lizards, birds' eggs, dung beetles, frogs and small mammals. They in turn were seized by the big martial eagles – proud ermine-chested hunters who wore the black hood of the executioner and were strong enough to carry off small antelopes.

Every day fresh hordes of insects hatched – wasps, bees, flies, ants, beetles, cicadas, grasshoppers – to be pursued through the grass by chanting goshawks, snapped up in mid-air by lilac-breasted rollers, or

. . . the big martial eagles – proud ermine-chested hunters who wore the black hood of the executioner . . .

winkled from tree-bark by wood hoopoes and scimitar-bills.

Every night the Marsh and the woodlands rang to the deafening choirs of tree-frogs, toads and bullfrogs – a cacophony of rattling, piping, clinks and whistles that continued intermittently until dawn. Owls hunted wide-eyed through the starlight. Bats and nightjars swooped endlessly through the shadows in pursuit of fat furry moths, sometimes surprising the silent form of the wall-eyed leopard on one of his nocturnal forays.

Slowly, the floods subsided. Although its muddy waters were still high, the Mara River resumed its old lethargic pace and the hippos returned to their favourite pools where they floated almost entirely submerged, blowing great watery sighs through distended rose-pink nostrils.

Where the rains had begun in earnest, the Talek lioness had moved her twins to the high ground overlooking the northern end of the Marsh, gently carrying them one at a time by the neck to a safer lying-up place among the whistling thorns. Three times in their first month they endured this indignity, dangling passively from their mother's mouth as she moved them from thicket to thicket. In a permanent hiding place they would have been too much at risk from marauding leopards and hyaenas.

OPPOSITE: They endured this indignity, dangling passively from their mother's mouth . . .

Their eyes opened after ten days and at three weeks they had lost the opaque milky hue of the newly-born. They had cut their first incisors and had learned to walk, if unsteadily at first, wobbling around on paws

which seemed too big for their fluffy spotted bodies. Soon they would be old enough to be introduced to the other lionesses in the group, but for the moment the Talek mother was content to lie in the grass and suckle them.

From her vantage point she could see the whole of the Marsh and all its comings and goings. Below her, water whelmed out of black volcanic rocks to mingle with the receding floodwaters and form a chain of shallow lagoons in which dense reed forests bristled like Maasai spears. This was the spring which, even in the severest droughts, never ran dry. After the rains, there was water everywhere, and the game dispersed throughout the Mara. But in three months' time when the wildebeest arrived, the plains would have dried. Then once again the Marsh would become an oasis for the thirsty herds, and the bubbling spring water would become their life-blood.

At all times of the year the Marsh was alive with the sounds and movements of birds, for whom the muddy waters provided a never-ending feast of snails, fish, frogs, dragonflies and other insects. Squacco herons skulked in the reeds. Sandpipers bobbed at the water's edge, where painted snipe probed for worms in the soft mud. Malachite kingfishers swung on the papyrus stems, blue and glittering like electric sparks, and armies of solemn storks – yellow-bills, saddle-bills, woolly-necks – slow-marched through the pools, stabbing at catfish and tiny jewelled reed-frogs.

Although competition was fierce, the Marsh was able to sustain large numbers of birds because many species had developed particular feeding skills. Thus the hammerkops – dwarfish brown herons with grotesque heads – paddled about in the shallows with a curious shuffling motion, stirring the muddy bottom with their feet to flush out small fish and tadpoles. The yellow-bills sometimes used a similar technique for catfish, but with an added refinement of their own. A stork would stand in the water with the tip of its partly-open bill submerged. Then, as it stirred the bottom with one foot, it would flash open a wing – a sudden movement designed to drive any disturbed fish straight into the waiting beak.

From dawn to dusk the great fen echoed to the incessant clangour of wild voices; the urgent chuntering of Egyptian geese, the mournful bugling of hadada ibis who cried their own name – *ha-da-da* – as they circled overhead with down-curved beaks outstretched. Sometimes the dark shape of a harrier or fish eagle would drift across the Marsh, or a marsh mongoose would come stealing through the sedgy tussocks. Then the reed-beds would erupt with sky-rocketing wildfowl and the air would fill with a cloud of wings, only to subside moments later, falling like a soft blanket across another part of the swamp.

To the right of the suckling lioness grew a wood of wild olives. Once it had been as dark and secretive as Kampi ya Chui, but the one-tusked matriarch and her destructive family had gathered there and knocked down many trees, letting in the light.

A similar fate had overtaken the giant acacia whose silver-grey carcass lay on the grassy shores of the Marsh not far from the spring. The tree had fallen many years ago. Too many elephants had pushed against its trunk, trying to shake down the succulent seed pods, until one day it had come loose from its roots and had crashed to the ground, smashing the eggs of a tawny eagle which had been nesting in the canopy. Now, stripped of its bark, weathered and polished by wind and sun, it lay at the water's edge like a discarded antler whose crooked tines offered a convenient perch for rollers, vultures, marabous and hammerkops.

Beyond the spring the water trickled for more than a mile, spreading out through tasselled reed-jungles of sedges and lily-ponds, before tailing off into a shallow depression and dribbling through a culvert under the dirt road between Governor's Camp and the airstrip. The northern half of the Marsh lay open on one side to the long swell of the plains, but the southern end was partly enclosed by trees reaching out from the riverine woodlands, forming an immense natural amphitheatre, home of a male waterbuck known for his record-size horns. Where the woods petered out on the far side of the Marsh there stood two venerable figs and a large acacia. It was under one of these that the Talek mother gave birth, and it was here, too, that Brando, Scar and Mkubwa were often to be found sleeping during the day.

On the other side of the Marsh rose the forest, its sombre walls following the twists and turns of the Mara River down to the distant Serengeti. Seen from the plains, the edge of the forest gave the impression of a green coastline, holding back the surge of grass. In places it thrust out long capes and promontories, islands of trees, between which were held deep bays and coves of wild meadow-like country where baboons foraged and waterbuck wandered among the domes of old termite mounds. The cliffs of the forest were the trunks of the tall trees: the fig, the wild olive and the African greenheart, softly mottled with silver-grey lichens; and in between the trunks, dark caves of shadow in which the gleam of a tusk, the slow flap of a sail-like ear, betrayed the presence of an elephant.

Life in the forest was very different from life on the plains. Here everything was secretive, surrounded, walled in with mysterious gloom. Even the sounds of the forest were different, and the creatures that uttered them seldom seen. From all sides there came the shrill whine of cicadas, which ceased only when the sun went behind a cloud; then, in the silence, the sudden bark of a bushbuck, the low belly rumble of a browsing elephant, the liquid voices of forest orioles would dispel the illusion that the forest was deserted.

The trees grew tall and straight – some of them to a hundred feet or more – meeting to form an airy canopy which cast a delicious dappled light on the tangled understorey of bushes, creepers and coarse forest grasses. The wind sighed as it passed over the forest, blowing the bottle-green leaves to silver; but at ground level the air was still and sluggish, heavy with the hothouse smells of humus, of centuries of leaf-fall, of trees growing, rotting and regenerating. In places there were clearings, bright pools of green sunlight among the deeper shadows and cobwebbed mossy trunks; and everywhere the winding trails of elephants, buffalo and hippos, a muddy palimpsest of printed hooves, strewn with piles of fresh dung whose moisture attracted clusters of swallowtail butterflies. And always, in the midst of the forest, the river, flowing slowly and silently between its steep clay banks. Along certain stretches the woodland thinned to a mere ribbon of bankside foliage, but in the vicinity of Governor's Camp it became a true forest, spreading out for at least four hundred yards on both sides of the river.

Here, then, in the low-lying grasslands around the Marsh, and in the park-like meadows between the Marsh and the forest, was the heart of the territory which the Miti Mbili pride had left and which Scar and Brando and Mkubwa would soon share with the three Marsh sisters and the Talek twins. Although they often drifted out across Topi Plain, or hung around the flooded murram pits where gravel had been dug for the Governor's Camp airstrip, the Marsh was becoming the epicentre of their world.

4

February-July 1978

The Great Migration

Every year the vast wildebeest nations of the Serengeti congregated here . . .

FAR TO the south, where the Gol Kopjes rose like ruined castles from the short-grass plains of the Serengeti, a female wildebeest chomped methodically at the close-knit turf. Only her hindquarters, deeply scarred where a lioness had once almost succeeded in clawing her down, distinguished her from her companions, a harem of perhaps two dozen wildebeest cows, many of them accompanied by their calves, gathered together under the watchful eye of a mature breeding bull. It was a pattern which repeated itself across the plains in every direction as far as the eye could see.

Every year the vast wildebeest nations of the Serengeti congregated here towards the year's end, and the scarred female had come with them. Mated the previous May, she had returned to give birth to her calf in the south-eastern corner of the park. The soils which carpeted the floor of the short-grass plains had been laid down in prehistoric times and were quite

different from the black cotton clays of the Mara. Eruptions of ash spewed out by the long-dead volcanoes of Ngorongoro had enriched the earth with calcium and phosphorus, minerals essential for healthy bone growth in the new-born calves, and if anywhere could be described as the true home of the restless wildebeest, it was here on these ancestral calving grounds.

Mother and calf grazed peacefully side by side. They ate in the manner of cattle, gathering the leaves of grass with the tongue and pulling it across the single set of incisors. The youngster was light brown, a sturdy bull-calf with long gangling legs, and was one of nearly half a million wildebeest born that year.

For nearly two decades the Serengeti wildebeest had been enjoying an unprecedented population explosion. Twenty years ago, there had been perhaps 250,000. Now there were nearly two million, and together with a million gazelles and a quarter of a million zebra, they formed the greatest concentration of wild animals left on earth.

Nevertheless, of the half-million calves now scattered across the plains, fewer than one in three would survive to adulthood. Many would simply lose their mothers in the densely packed sea of animals, and fall easy prey to predators. Often a new-born calf would become separated from its mother before it had learned to recognize her. Then the infant would attach itself to anything that moved, sometimes even tottering hopefully towards a waiting lion.

When the great migration began, many more youngsters would be swept away and drowned during the massed river crossings, dragged down by crocodiles or trampled beneath the frantic press of bodies from which others would emerge hobbling on broken limbs to await the inevitable hyaenas. And always there were the ravenous hunters: the packs of brindled wild dogs whose sudden appearance, loping over the dawn horizons, threw the herds into headlong panic; the rangy cheetahs with their devastating stride; and the ever-watchful prides of lions through whose territories the wildebeest would have to pass.

From the very beginning, every year in the life of the young bull-calf would be a game of chance, a never-ending obstacle race which only the fittest and most fortunate could hope to complete. His greatest hope of survival lay in the sheer numbers of calves born on the plains during January and February. Of all the mysteries interwoven in the social behaviour of the wildebeest, none was so remarkable as the way in which the vast majority of cows managed to give birth within two or three weeks of each other. It was almost as if the swarming herds possessed a communal womb. Furthermore, each mother-to-be was able to control the moment of birth, holding back if danger threatened, and usually waiting until morning, when most of the meat-eaters were already gorged and only the jackals and tawny eagles were still around to feast upon the after-births.

There was no hiding place on the treeless plains, no shelter from hungry eyes. The calf had been born in the open, and, although he had been able to stand within five minutes of struggling from the foetal sac, he was utterly vulnerable. Only the great flood of births, the bleating presence of thousands of other infants of his own age, each one of whom might be eaten instead of him, increased his own chance of survival. Without the natural strategy of mass births he would have been an easy target for carnivores. But as one of such a vast crowd he was totally inconspicuous; and the herd instinct, the comforting feeling of safety in numbers, would govern his wanderings for the rest of his life.

Immediately after the rains the Serengeti pasture was so prolific that a single acre might produce eighteen tons of protein-rich fodder. But by the end of April the verdant plains had already begun to revert to yellowing stubble. Soon the time would come when they would no longer be able to sustain the herds; and then, as if in obedience to some inaudible command, the migration would begin. The zebra and wildebeest would

stream out of the short-grass country, heading for the taller grasses and open woodlands of the Western Corridor.

In the meantime, mother and calf ate steadily. They rested little, sometimes grazing up to eighteen hours a day, storing energy for the long march ahead. All around them were similar groups, some with mothers and calves, others composed entirely of sub-adult bachelors who would one day challenge the mature bulls and try to win harems of their own. Already the rut was beginning as the rich grazing brought the animals into peak condition, and battles were breaking out everywhere as the territorial bulls sought to round up and defend groups of females.

The wildebeest were never silent. Even when feeding they muttered and grunted, and the mumble of their voices carried far across the plains in a resonant drone, like a giant hive. Now, as the rut began in earnest, the noise grew even louder, with the frenzied bulls adding a drum roll of hooves as they galloped endlessly around their territories.

With their long, lugubrious faces, their bearded chins and fly-whisk tails, wildebeest are the buffoons of the plains, absurd and ill-proportioned. Their heads are too big and their hindquarters too small. Their rickety legs appear too puny to support the weight of their bodies. They have the horns of an ox, the mane of a horse and the sloping back of a hyaena. They are a grotesque joke, a parody of the antelope family to

He had been able to stand within five minutes of struggling from the foetal sac . . .

which they belong; yet they are the most successful of all the animals in the savanna, and their immense journeys, cropping the turf like locusts as they spread across the Serengeti, are essential for the good of the grassland.

Scattered across the plains were docile groups of male wildebeest, but as mating fever swept through the herds the sexually mature animals left the bachelor troops to seek individual territories and win harems for themselves. At the centre of each territory was a bare patch of ground where each bull spent most of its time. These were their stamping grounds, and the grass was soon worn away by the constant pawing, kneeling and horning of their courtship ritual. Overnight they became hyperactive and truculent. Sometimes they rubbed their faces on the ground, or rolled in their own dung and urine. At other times they advertised their presence by standing for long periods with heads held high, staring into the distance as if watching for predators.

If other wildebeest entered their territories, the bulls would swish their tails and canter out to meet them. Females would be greeted with ears held down in a sexual display of welcome, but bachelor bulls would be vigorously expelled. The oddest encounters were those which took place when one territorial bull strayed into the area of another. This was the signal for an extraordinary ritual challenge. The behaviour might take several forms. Sometimes one bull would present himself sideways-on to the other, as if to intimidate him by sheer size, or they would stand end to end, rubbing their heads on each other's rumps. If one feigned a charge, both animals would drop to their knees in the classic battle position of the rutting wildebeest. But instead of clashing head-on they would hook viciously at the grass, tearing clods of earth out of the ground; then they would jump up with a snort and begin to plunge and cavort like rodeo stallions. In the end their aggression would subside as suddenly as it had begun, as one or other lost interest and wandered off.

Both animals would drop to their knees in the classic battle position of the rutting wildebeest . . .

The full moon brought the rut to its climax by inducing the vast majority of cows to come into season simultaneously. This was the secret which lay behind the wildebeest's mysterious ability to give birth within a few weeks of each other. Mass impregnation now would repeat the miracle of half a million new-born calves the following February. Many bulls were so overcome that they foamed at the mouth, and even the bachelors were convulsed by sudden pelvic spasms that brought them to the point of ejaculation.

As the rut progressed, the plains presented a scene of bewildering chaos, with thousands of wildebeest milling about in grunting masses, and the breeding bulls endlessly circling, galloping, challenging and mating. To add to the confusion, the migration started while mating was still at its peak, and huge numbers of animals began to depart in long straggling columns. Among them were the scarred cow and her calf.

So strong was the territorial imperative that many bulls remained on their stamping grounds long after the females had deserted them; but in

Huge numbers of animals began to depart in long straggling columns . . .

the end they, too, felt the irresistible pull of the migration and were sucked into the westward-flowing tide. Soon only a scattering of wildebeest remained on the spent grasslands, spinning like dervishes in a ghastly dance of death. These were animals which had been stricken by a small fly which laid its eggs in the wildebeest's nostrils. The eggs hatched into maggots which bored their way into the brain, and an infected animal would dance until it dropped.

A pall of dust hung over the moving herds. Some went slowly, heads nodding to the rhythm of their gait; others thundered past at the gallop, with manes and tails streaming. Behind them they left a maze of narrow trails, where their trampling hooves had cut deep grooves in the skin of the land. The routes they followed were seldom those of the previous year. No two migrations were alike. Only the general westward drift was predictable, for their movements were ordained by the retreating rains. Sometimes the herds would double back when a rogue storm produced a flush of new growth behind them, but the main thrust of their exodus was irreversible.

Where kopjes pierced the infinity of plain and sky, the herds parted like swirling currents and flowed past on either side. The kopjes are very old. Long before the first wildebeest appeared on earth, their granite domes had been smoothed and wind-worn into strange elephantine shapes. Grey and weather-stained, marooned in the illimitable wastes of drying stubble, each rocky outcrop is a closed and secret world of its own, a watch-tower, a food source and a sanctuary for many different creatures. Lions and leopards nurse their cubs among the rocks. The hyrax dwells there, and the agama lizard. When the moon rises, spitting cobras emerge from their holes to prey on the scuttling agamids, and troops of dog-faced baboons bark from their cliff dormitories. At sunrise the baboons descend and spread out into the plain, foraging up to fifteen miles a day in search of roots, bulbs and scorpions; but always they return before nightfall.

The kopjes are not entirely barren. Fig trees sprout from rocky fissures, and frosty-blue aloes with fleshy leaves. Some kopjes have permanent springs, or hollows that hold the rain, attracting small terrapins which survive the long droughts by burrowing deep into the mud. During the day, klipspringers stand like statues on the kopjes for hours on end, watching for predators from their dizzy ledges. They are antelope relatives of the wildebeest; but, unlike the restless and far-ranging herds, their entire lives are anchored to the same rocky island.

The sure-footed klipspringers share their crags with peregrines and lanner falcons – hunters of the gentle doves – which soar about the cliffs with harsh heckling cries, or sit unmoving with bulging crops on favoured ledges splashed white with their droppings. Rarer are the Verreaux's eagles, which build their eyries in the Moru Kopjes beyond the Mbalageti River. The eagles kill and eat the hyraxes which bask on the rocks every morning, staring unflinchingly into the sun. Surprise is their most effective weapon. Patrolling the kopjes on long black wings, they sweep round a corner and seize a hyrax in their big yellow fists before it has had a

The hyraxes bask on the rocks every morning . . .

chance to dive for cover. They are powerful birds, strong enough to carry off a young baboon or knock a klipspringer from the rocks, and even the leopards are wary of them.

In the treeless void of the short-grass plains, only the kopjes offered shelter from the sun. But the wildebeest shunned the beckoning islands. Fearful of the predators their shadows might conceal, they plunged on in endless columns over the rim of the western horizon, and the klipspringers watched them go. It would be November before they would see them again.

*   *   *

The wildebeest mother, tramping along with the bull-calf at her heels, was no stranger to the migration. This was the seventh time in her life that she had undertaken the annual trek to the summer pastures of the Mara, and she had survived all its perils and pitfalls, the droughts and river crossings, the poachers' snares and the lions in the woodlands. But for the youngster it was the first and most dangerous time. Ahead of him stretched a daunting marathon, a round trip of some three hundred miles – if he survived.

Bands of zebra moved into the tall grass ahead of them, mowing down the meadows with their sharp cutting teeth. Unlike the antelopes, they could cope with the tough fibrous stems of the taller grasses, leaving the choice green blades of oat grass for the more fastidious wildebeest. With them travelled other herbivores, a spectacular concourse of semi-migrating hartebeest, Thomson's and Grant's gazelles. In all, there were at least 200,000 zebra, and perhaps a million gazelles.

Whenever the wildebeest moved into a new area and stopped to feed, the breeding bulls tried to establish temporary territories and round up groups of females, but as soon as the grazing was exhausted the herds would move on, marching nose to tail in long trains which sometimes stretched to the horizon. Even when panicked into headlong flight they would go to almost any lengths to keep the line intact. Where predators lurked it was always safer to follow in the footsteps of others rather than to wander at random and risk an ambush. Over the millennia, natural selection has favoured animals which follow dumbly rather than those which stray and the habit has become embedded in the psyche of the herds.

By the beginning of June, mother and calf were moving with the mainstream of the migration through the Seronera Valley. Magnificent stands of fever trees grew in airy glades beside the Seronera River, shading the banks with their flat-roofed canopies of lacy green. The wildebeest avoided the pleasant coolness of the glades, for the trees were a favourite haunt of leopards, whose spotted coats were almost invisible against the mottled yellow bark.

There were lions in the valley, and cheetahs with hungry cubs. So the herds kept to the treeless places as much as possible. Only in the early morning, and again in late afternoon, did they feel compelled to drink from the river. Then they approached warily, the leading animals stamping and snorting with fear as they passed among the trees, for their senses told them that here was a place where danger watched from the shadows.

On their second day in the valley, the cow and calf found themselves at the forefront of a group which had spent the day grazing far out on the open grasslands. Now they were coming in to drink, led by a bull who had completed the migration at least half a dozen times. Immediately behind him was a mother with a yearling female; then came the cow and calf. As always the bull was cautious; he had seen too many of his companions pulled down by cats or drowned by crocodiles. As he approached the river, the deceptive tranquillity of the fever glades filled him with a sense of unease that deepened when he saw a pair of giraffes staring fixedly into the trees. The bull stopped abruptly, recognizing their behaviour as a signal that predators might be lurking. For several seconds he froze, straining his eyes to see into the shadowy tangle of fallen branches and rank matted grasses. But his hoarse snort of alarm came too late to save the yearling. Even as he whirled, the grass exploded past him in a wicked yellow blur.

Later, when the sounds of stampede had faded, the leopard ate the liver and part of the stomach of the yearling wildebeest, then took the still-warm carcass in his jaws and sprang into one of the fever trees, where he wedged it securely in a high fork out of the reach of hyaenas.

*   *   *

In the confusion that followed the leopard's ambush, the bull-calf became separated from his mother and ran bleating among the milling herds. Although he no longer needed to suckle and could fend for himself, his

Herds of impala stood quietly at the edge of the plains . . .

chances of survival without her would be slim. Luckily he found her before dark and the pair of them settled down to graze as if nothing had happened.

The following week found them on the Musabi Plain, deep in the Western Corridor, where they stayed to feed until the grass was exhausted. Then, as the migration recovered its momentum, they swung north in a swirling clockwise advance across the Grumeti River. Many wildebeest perished in the crossing and the Grumeti crocodiles grew fat on the bodies trampled in the mud, but the cow and calf were not among them. They cantered north after the zebra, running the gauntlet of poachers' snares and arrows as they left the park and poured through the bush country behind Fort Ikoma. Re-entering the park beyond the Baracharuki Falls, they continued northwards through a changing landscape of dry rocky hills, water-courses and belts of commiphora scrub. Here the grasses still grew tall and thick, but the main feature of the country was the ubiquitous acacia, the African thorn tree whose familiar, flat-topped crown provided shade and shelter for the animals. The trees seldom grew close enough to form a continuous closed canopy, but were widely spaced, with glades and meadows that gave these northern Serengeti woodlands a pleasing, park-like appearance.

The grassy woodlands held their own resident populations. Warthogs glowered from their burrows. Timid dik-diks, antelopes scarcely bigger than hares, stared from their miniature territories under the *ngoja kidogo* – the wait-a-bit thorns. Herds of impala stood quietly at the edge of the plains, and giraffes moved solemnly through the glades. Leopards slept in the giant fig trees that overhung the swampy pools of drying water-courses, and buffalo wallowed in the reeds, their presence betrayed by a sudden flutter of egrets. Everywhere, smashed and fallen trees, with bark and branches freshly ripped, showed where elephants had browsed during the night. When daylight came they had moved off into the grizzled thornscapes of commiphora, for ivory poachers had made them wary of the open places, but their glistening dung was everywhere. They were the architects of the wide savannas, path-makers, forest clearers, diggers of wallows and water-holes, constantly modifying the landscape. Yet in spite of their bulk, their destructive feeding habits and noisy trumpetings when disturbed, they could move as silently as wraiths, for they walked on their toes, each massive foot encased in a sack of skin which deadened all sound.

All day long the woodlands rang with cries of barbets and bush shrikes, hornbills and laughing doves. Guinea-fowl cackled from sandy dongas, coucals bubbled in the thickets, and even when the birds fell silent the glades were never entirely still. Baboons barked. Zebra yelped. Carpenter bees droned through the long hot afternoons and cicadas sizzled the air with their incessant rasping din.

The passage of the migrating herds attracted swarms of tsetse flies. Denizens of the thickets and woodlands, the tsetses could not abide the brightness of the open plains, but sought the shady undersides of branches overhanging the game trails. There they clung motionless, scissor-wings folded over bullet-grey bodies, almost invisible except for the hot red glint of their compound eyes. Warthogs were their favourite prey, but the tsetses were not fussy. They would feed indiscriminately on the blood of wildebeest and zebra, eland, kongoni, lions, hyaenas, human beings – and even lizards. When their finely-tuned sensors detected the presence of moving animals they flew out from the bushes in furious biting hordes, settling on flanks and bellies where skin was thinner, expertly stabbing with their needle-sharp probes. Afterwards, too full to move, their wine-dark abdomens gorged with blood, they stuck to their hosts until they were able to digest their meal and lumber away.

The tsetses are carriers of trypanosomes – parasites transmitted by their bites which cause sleeping sickness in men and cattle – but most of the wild ungulates have long since become immune. To them, the flies are merely a nuisance – though at times their persistence so maddens the wildebeest that they break into a mad gallop to try and escape. Only the drongoes – sooty fly-catching birds with forked tails and an upright perching stance – welcome the tsetses, cracking their hard bodies in sharp hooked bills.

As the dry season advanced, kongoni and zebra came down from the higher ground where the grazing was finished to join the wildebeest herds in the crackling woodlands. But even here the rank hay was soon consumed and reduced to stubble, so great was the army of hungry animals. Soon once again the herds were marching north, heading for the Mara. With them went the cow and calf, relieved to be leaving the tsetse-infested woodlands. The trampling throng began to move with increasing urgency. They travelled day and night, sometimes careering along in a wild gallop when the leaders took fright. On either side of them were other columns, intermingled with bands of zebra and smaller groups of resident kongoni. Large numbers of gazelles had followed them, too, but they would not make the great river crossings into the Mara, preferring to remain in the Serengeti woodlands until the rains returned. And behind them all came the stragglers, a pitiful procession of the old, the lame, the sick, the weak, the luckless calves who had lost their mothers. None of

them would last long. Vultures monitored their progress from the skies, and always, moving at the edges of the migrating herds, flitted the spectral shapes of the ever-hungry carnivores.

. . . a growing movement of wildebeest converging on a wide bend of the Mara River . . .

So far the cow and calf had been lucky. They had managed to avoid the ambushes of lions and leopards in the acacia woodlands. They had survived the long treks across the open grasslands, where hundreds of their companions had fallen to cheetahs, hyaenas and hunting dogs. They had come through the perilous river crossings unscathed, though the Grumeti and the Bologonja Rivers had both claimed a host of drowned bodies for the crocodiles.

By the end of June they had crossed the Sand River into Kenya and headed north past Keekorok Lodge to the Burrungat Plain. Here they were joined by the resident Kenyan herds, returning to the Mara from the Loita Plains. Ahead of them raced a cavalcade of zebras. The horses were always the first to appear in any new area, moving eagerly into the tall meadows to feed on the waving red-oat grass; but the wildebeest were never far behind, and the Ruppell's griffon, swinging in his lofty station five thousand feet above the herds, saw them swarming like ants into the lush dry-season pastures of the Mara Triangle. There they remained for most of July, resting and feeding after their long ordeal, while every day new arrivals swelled their numbers until the plains were black with animals. Towards the end of the month, much of the area had been grazed to the ground, and the cow and calf found themselves among a growing movement of zebra and wildebeest converging on a wide bend of the Mara River below the Serena Lodge.

* * *

The main crossing had begun several days earlier, and by the time the cow and calf reached the river they were confronted by a scene of indescribable confusion, with masses of animals milling about on both sides. The herds that had managed to cross were now bustling away into the long grass of Paradise Plain, but hundreds more animals stayed behind on the banks, as if reluctant to leave until they had been joined by those still waiting on the other side.

The noise was deafening. Rising tension transformed the muttering herds into monstrous grunting choirs, the wildebeest's voices pierced by the barking of zebras and the bleating of lost calves. The air reeked of animals, of mud and dung and trampled grass. Everywhere flies buzzed and crawled amongst the hot press of bodies, gathering in damp clusters around eyes and nostrils; and as more and more wildebeest arrived to join those waiting to cross, clouds of dust billowed into the sky, obscuring the sun.

For the scavengers, the dust clouds were the sign of carrion. Already the stench of rotting flesh hung about the river, where bloated bodies lay piled against the banks or snagged against half-submerged branches – casualties of earlier crossings – and the vultures had been busy. Many of the corpses were spattered with white droppings where the birds had crouched to rip out the eyes. Now the vultures gathered in dark huddles on the banks, or filled the trees with their macabre silhouettes; for the eyes and tongues were all they could scavenge until the pent-up gases burst the distended bellies, or until the river softened the carcasses. Even so, some birds were so ravenous that they would alight on a floating body and sail downstream, balancing with half-crooked wings as they plunged their heads under the water in desperate attempts to reach eyes, tongues, sphincters – any point of ingress that might enable them to pluck at the putrid treasures within.

The crocodiles did not need to wait. Attracted by the turmoil in the water, they would slide silently from banks and sand-bars, nosing stealthily upstream with only their eyes above the surface. The wildebeest and zebra seldom saw them coming. In their haste to cross the river they never noticed the approaching arrow-ripple of a saurian snout, or the swirl in their midst as a victim was dragged under. Now the river was so full of drowned animals that the crocodiles did not even have to kill their prey. At night they gathered in the shallows where the bodies lay in drifts against the rocks, grasping the limbs in their jaws and spinning their bodies to tear away chunks of flesh.

Upstream from the crocodiles, more carnage marked the spot where another big crossing had taken place. Here the banks had been churned into a muddy quagmire from which protruded the heads and limbs of dead and dying animals. The easiest exit points from the river were through the narrow gullies trodden by the hippos; but, as the hippo trails became choked with corpses, the animals still struggling in the river tried to scramble up the steep banks. Desperation drove them on, but many fell back to crush those waiting beneath, or themselves became trampled underfoot. At one point a young wildebeest, apparently resting at the water's edge, suddenly struggled to his feet, only to reveal a smashed hind leg inextricably wedged in the roots of a submerged tree stump. Beside him sprawled another dozen animals which had tried in vain to free themselves from the morass and were now too exhausted to move. Above them towered the bank, surmounted by a wall of animals. So great was the compulsion driving the herds forward that the wildebeest on the edge were forced to leap into the water, falling fifteen feet or more with their forelegs tucked under them. The lucky ones plunged into deep pools and were able to swim to safety on the other side; others, less fortunate, crashed on rocks or tree stumps.

At one point, a section of overhanging bank collapsed under the weight of the herd, burying the animals trapped beneath it in an avalanche of mud and bodies. By the time the pandemonium had subsided, the river had claimed another score of victims.

The zebra did not falter . . .

*The multitude advanced,
leaping and crashing into
the river . . .*

. . . a scene of indescribable confusion . . .

He saw a hippo trail, and raced into it . . .

*The hippo trails became
choked with corpses . . .*

*Bloated bodies lay piled
against the banks . . .*

*They would alight on a
floating body and sail
downstream . . .*

*The crocodiles did not even
have to kill their prey . . .*

*Afterwards, in the lull, small groups of topi crossed
the river, followed by a group of buffalo . . .*

Afterwards, in the lull, small groups of topi crossed the river, followed by a group of buffalo. It was now midday, and still the cow and calf had not yet entered the river. But, as the shadows lengthened through the afternoon, the herds began to push forward again. As often happened, it began with the zebra. At first they came cautiously, snorting and staring at the water's edge, only to wheel round and rush back up the bank from some real or imagined threat. But when other zebra on the other side of the river began to bark at them, the stallions answered with excited squeals. This time they did not falter. Urged on by the barking of their companions, they broke into a wild cavalry charge and plunged three abreast into the water. For a moment, the massed ranks of wildebeest had seemed mesmerized; but the sight of the zebra once more jolted them into action. Again the dust clouds boiled into the sky as the multitude advanced, leaping and crashing into the river.

Wedged in the middle of the main wave, the calf cried out as his mother disappeared in the flying spray. Caught up in the chaos of plunging bodies, he was carried helplessly forward, aware of nothing but the frantic expressions of the animals around him, of lips laid back and wild rolling eyes as the desperate animals struggled to keep their heads above water. Eventually he felt rocks beneath him. Somehow he managed to pull clear of the mud which sucked at his hocks in the shallows. Ahead of him he saw a hippo trail, and raced into it on the heels of a flying column of zebra to emerge gasping but unscathed on the edge of Paradise Plain. All around him wandered dozens of other gangling youngsters, bewildered orphans

. . . picking his way over the maze of hoof-prints . . .

like himself. The main body of wildebeest had not waited for the stragglers, but raced on towards Rhino Ridge as if determined to rid themselves of the nightmare of the crossing. The long lines of cantering shapes vanished into veils of dust, and in the end the calf trotted after them.

Now the sun falling swiftly towards the escarpment lit the grasslands with an unearthly saffron glow. This was the golden hour, the heat fading, the diurnal birds falling silent until only the tiny cisticolas still called among the untrampled swales of Rhodes grass. Over the horizon wandered the wildebeest calf, a lonely silhouette briefly imprinted against the ball of the sun as it touched the earth. In the thickening dusk, scops owls began to trill. Later, from farther off, came a more sinister sound as hyaenas whooped to each other in sepulchral voice, and towards midnight the roars of the Paradise pride rolled across the plains.

The Paradise lions roared again just before dawn, but by this time the calf had left their territory and was crossing Topi Plain. During the night he had encountered the Murram Pit hyaenas hunting by starlight on the Governor's Camp airstrip. Their hideous cries and dim slouching shapes filled him with dread, but they did not chase him when he ran.

Now, stumbling from exhaustion, he could hear the growing mumble of the herds, see the glitter of their horns in the early morning light as they moved through the bushes on the ridge above the Marsh. Wearily he skirted the edges of the reed-beds, picking his way over the maze of hoof-prints left in the mud where the wildebeest had been drinking. He had forgotten his mother. His sole instinct was to rejoin the herd. There, surrounded by the soothing presence of his own kind, he might rest and feed and feel secure once more. Barely a hundred yards now separated him from the slowly moving herds. Summoning up his last reserves of strength, the calf trotted towards them. As he did so, a flat tawny head rose briefly from the bending grasses, and just as swiftly sank from sight.

5

July-September 1978

The Murram Pit Clan

THE MARSH lioness lay on a termite hill, absorbing the warmth of the early morning sun. Beside her sprawled her two sisters and the barren Talek twin. All four stretched out on their backs, eyes half closed, white bellies still slightly swollen from the warthog they had eaten the previous day. They paid no attention to the grunting of the wildebeest herds on the nearby ridge but, when the lone calf came blundering towards them, the Marsh lioness was instantly alert. Soundlessly she slid from the mound and hid herself in the grass. Her companions remained where they were, but they, too, were now wide awake and watching.

The crouching lioness trembled. Her hindquarters bunched like a coiled spring as she dug her claws into the ground, then she launched herself in a series of low bounds. The calf shot away, jinking in a vain attempt to avoid her, but the lioness turned easily in mid-stride and slapped him down with her paws. Before he could rise she had him by the throat, and held him until the other lionesses arrived.

There now began a macabre game of cat and mouse with all four lionesses joining in. Every time the wildebeest tried to escape it was bowled over by one of the playful females. None of them was especially hungry, but eventually, when they tired of their sport, the Talek twin closed her muzzle over his windpipe until his limbs ceased their twitching. Then they settled down to feed half-heartedly.

The calf was the first of many wildebeest to be killed by the Marsh carnivores that year. For the Marsh lions and all the Mara prides and predators, the migration was a movable feast that promised easy hunting and full bellies until the herds departed in October. Sudden death is no stranger on the plains. Nature has no favourites. Wildebeest die. Lions die. But the wildebeest can be counted in millions, the lions only in hundreds, and while the loss of even several thousand calves scarcely affects the

She launched herself in a series of low bounds . . .

The sight of the huge male sent them scurrying back to their mother . . .

strength of the herds, the food was welcome that morning to the lionesses whose unborn cubs were kicking inside them with growing vigour.

One by one, when their time came, the three Marsh sisters retreated into the forest at the edge of the reed-beds. There, in secret thickets and among the knotted roots of a forest fig, they produced the cubs sired three and a half months earlier by Brando, Scar and Mkubwa. There were eight young in all. Two of the sisters had triplets; the other bore four cubs, but only two were alive.

For all three lionesses it was their first experience of motherhood and, as often happens in a pride, they gave birth within two or three weeks of each other. For the first six weeks of their lives the helpless cubs were kept well hidden. Concealed in their forest nursery, they slept and suckled and slept again. But, as the days passed, they became more playful and adventurous, wrestling and tumbling amongst themselves and biting their mothers' tails with their milk teeth.

Even when the cubs were still quite small, the Marsh sisters resumed their old close relationship. Often they hunted together, stalking the wildebeest which came trooping off the plains to drink in the reeds. Sometimes they stayed together all day, lolling in the shade until dusk sent them padding anxiously in search of their young. If a lioness approached one of her sisters as she lay suckling her cubs in the forest, there was no hostility; but when Scar arrived one morning to find the cubs feeding at a kill, the sight of the huge male sent them scurrying back to their mother.

The Marsh lions were not the only pride with cubs. In July, Young Girl had detached herself from the Miti Mbili group to give birth to Brando's second new family in a patch of dense scrub on the upper reaches of Miti Mbili lugga. She produced five cubs, but two of them were dead within a week, killed by the wall-eyed leopard. Young Girl had heard his sinister cough the previous night and had already begun to move her litter to another hiding place at daybreak, gently lifting them in her jaws and carrying them one at a time down the lugga. But no sooner had she left with the third cub than the old tom sniffed out the remaining pair and killed them.

Soon afterwards, an even greater tragedy overtook the Miti Mbili pride. For several days Old Girl had been limping badly from an encounter with a bull buffalo. The buffalo was clearly sick, which was why the hungry lionesses had attacked him, but he did not die easily. When the old matriarch had sprung at him he had wheeled like lightning and hooked her in the groin. The swing of his horns had bowled her over, leaving her winded, and he had caught her again under the ribs before Notch and Shadow managed to overpower him.

Such encounters are an everyday hazard for lions and the slightest misjudgement can be severely punished. During their brief and violent lives, they often receive appalling injuries as a result of fighting among themselves or cornering a buffalo; but they also possess miraculous powers of recovery, so that even quite large wounds seldom incapacitate them for long. Sometimes, though, a bad injury proves so debilitating that the body can no longer fight off the insidious effects of parasites. And sometimes, as in the case of Old Girl, a dirty horn wound will cause septicaemia.

As the infection spread through her groin, one hind leg swelled until she could only hobble. At first she kept herself alive by feeding on kills made by her pride companions, but towards the end, when she could no longer keep up with them or fight for her share, she dragged herself to the shelter of some rocks at the edge of the Miti Mbili Plain and lay down to die. The hyaenas found her there next day, but it was a long time before they dared to approach the huge grizzled body.

Death on the plains is a bloody business, yet the zebra who are disembowelled and eaten alive by hyaenas do not suffer as much as Old Girl did. Extinction in the jaws of a predator may be crude and brutal, but it is infinitely more merciful than a lingering end through disease or old age.

Another to suffer in this way was a large bull hippo, who came to the Marsh after the long rains. The big river-horses lived in the wide pools where the Mara flowed sluggishly between steep cliffs of mud. With their short legs and massive barrel bodies there is nothing equine about them except their name. They are really gigantic pigs who have become adapted to an aquatic life and, like the warthogs whose ancestry they share, their yawning jaws reveal a formidable armoury of crooked yellow tusks.

The bull was a huge beast, nearly three tons in weight and in the third

decade of his life. A veteran of many brawls, his muddy grey hide was criss-crossed with livid scars – battle honours received in countless quarrels over the schools of females. For many years he had fought with one rival after another, sometimes inflicting such terrible wounds that his opponents had bled to death. But in the end he had met his match in the shape of a younger male. For half the night the river had echoed to the crash and shock of their ponderous bodies as the two dreadnoughts charged at each other with infuriated bellows. Furiously they slashed and parried with gaping mouths until the old warrior felt his strength fading and was forced to abandon the river and lumber off into the safety of the Marsh. There he wallowed in the reeds, coating himself with a salve of soothing mud, but his wounds did not heal. Oxpeckers probed in the festering flesh, adding to his misery, for the fight had drained him to the point of exhaustion, lowering his resistance to disease.

He became a familiar and pathetic figure as he plodded listlessly among the reeds . . .

At night he still emerged from the sanctuary of the sedgy pools, travelling by the same well-worn paths to feed in the grassy meadows near the Marsh. Normally he was an avid browser – but as the sickness spread inside him he ate less and less. In the weeks that followed, he became a familiar and pathetic figure as he plodded listlessly among the reeds – yet, in spite of his weakness, he was as belligerent as ever. Sometimes the Marsh lionesses would meet him on his nightly wanderings, and the fierce *whoosh* of breath expelled through flared nostrils as he shook his head with rage was enough to keep them at a distance. When at last he died, the stench of his corpse hung in the reeds for days, and tawny eagles danced on his sprung ribs until the hyaenas moved in to dismantle him, leaving nothing but his skull in the water.

In the natural world of the savanna, the hyaenas are the great levellers. No animal is so fleet or so powerful that it might not one day be consumed

Tawny eagles danced on his sprung ribs . . .

There is nothing made of flesh and bone that they cannot devour . . .

by their voracious clans. They possess the most powerful predatory jaws in Africa. There is nothing made of flesh and bone that they cannot devour, few creatures they will not kill when the odds are right. Their reputation for scavenging the kills of others obscures their true nature. They are skilled and ruthless hunters; and, despite their ungainly appearance, their bodies are ideally shaped for the relentless running down of prey.

Although hyaenas can go for days without drinking, they delight in the presence of water, frequenting pools and luggas where they can sprawl on their briskets in the muddy shallows and escape the irritation of tsetses and other biting flies. In the Mara, during the heat of the day, they often use the edge of the Marsh as a resting-place, but they are more commonly seen around the murram pits south of Leopard Lugga. The pits, which had been dug to provide hard-core for the road to Governor's Camp, fill with water during the rains, and for most of the year they are a magnet for the hyaenas, who raise their cubs in two large dens at the southern end of the lugga.

The Murram Pit clan, as it became known, was a group of twenty adult hyaenas and almost as many juveniles of various ages. Throughout the migration they lived almost exclusively by hunting wildebeest, killing mostly between dusk and midnight, or in the early morning, when the sight of even a solitary hyaena shambling over the horizon, head high, tail up, round ears cocked forward in characteristic hunting pose, would provoke snorts of alarm from the watching herbivores.

To the north, the hunting grounds of the Murram Pit hyaenas extended only as far as Fig Tree Ridge, where they sometimes fought bloody inter-tribal skirmishes with the rival Gorge clan, but all the Marsh was theirs. When the hippo died in the Marsh it was the Murram Pit clan which demolished his ripe carcass, shearing through hide and sinew with their sharp carnassial teeth and cracking the big thigh bones with their powerful cone-shaped pre-molars to get at the marrow. And when one of the Marsh lion cubs disappeared towards the end of August, it was a Murram Pit hyaena that killed it.

The hyaena had discovered the hiding place at the base of the hollow tree where the lioness had left her two cubs, and had bitten one of them through the spine while the other cowered out of reach deep down beneath the tangled roots. The hyaena had not attempted to kill the second cub, fearing the lioness's return, but had carried off the dead one to eat at leisure.

The Marsh lions and the Murram Pit hyaenas were no strangers to each other. Since the lions had first moved into the area there had seldom been a night when their paths had not crossed on the moonlit grasslands of Topi Plain. They shared the same territory, stole each other's kills and regarded each other with a deep-rooted and justifiable suspicion. The lions, confident of their greater strength, ignored the hyaenas unless heavily outnumbered, but during the regular confrontations over kills their ill-disguised antipathy sometimes erupted into violence.

It was precisely such an encounter that had once almost cost the cub-killer her life. Normally the hyaenas were uncannily astute in judging the mood of their more powerful adversaries. Although they were bold enough to try and drive the young Marsh lionesses from a kill, advancing aggressively with manes and tails raised, they were usually prudent enough to leave the big males alone and wait at a safe distance until they had finished eating. But on this occasion the female had tried to steal from Mkubwa and had paid for her rashness. Her blunt, bear-like muzzle still carried the marks of his claws, and one side of her mouth had been so

The hyaenas raise their cubs in two large dens at the southern end of the lugga . . .

badly ripped that she was left with a permanent lop-sided grin. But afterwards, seemingly none the worse for her experience, she and her raggle-taggle followers continued to harass the Marsh lions as boldly as ever. She was an exceptionally big animal. Her scruffy coat had already begun to lose its spots, showing her to be well into her prime, and she had long enjoyed a high-ranking position in the matriarchal social structure of the clan.

Within every clan there are always some hyaenas which are fitter and faster than the rest – those which display more ferocity and are always at the forefront of the chase. These are the butchers, the natural leaders who are invariably among the first to be in at the death. The cub-killer was just such an animal. Her formidable size and insatiable appetite had made her an inveterate hunter, but when she killed it was not through any comradely feeling for the rest of the clan. She killed for herself, gulping down the biggest and choicest mouthfuls of flesh before the others could catch up with her.

The lion cub had not blunted her appetite for long. Shortly after dark she awoke, scratched, and began to groom herself, licking and biting at the parasites itching under her ginger fur. In her manner of grooming she was more like a cat, curling her body and raising a hind leg as she sat and licked at her genitals. Like all female hyaenas, her extraordinary elongated clitoris and false scrotum made it almost impossible for a human observer to distinguish her from the males. Only when in milk do the females become easily identifiable by the presence of prominent black nipples.

Grooming completed, she rose to her feet. With the coming of darkness her hunger had returned. A light breeze tugged at her tangled mane, and the air held the smell of gazelles and wildebeest, but she did not hurry. She walked alone, and as she nosed through the damp grass she whooped from time to time; a chilling sound that slurred deep in her throat, almost like a growl, rose eerily and died away. The moon sailed out from behind a cloud. The grasslands were a ghostly silver-grey. Again the mournful voice, rising, falling, fading, silence. Out on the plains, horned heads lifted in sharp silhouettes.

Uneasily the Thomson's gazelles watched the black shape approaching. Some fled, but the adult males were unwilling to desert their territories and stood their ground, banded flanks twitching uncontrollably as she prowled among them. She was looking for fawns which might be hiding among the tufts of nibbled grasses, but, finding none, she stopped and stared at one of the territorial males. At once the buck's nerve broke and he bounded away in the curious stiff-legged gait known as stotting or pronking, with the hyaena in pursuit. Stotting is often an effective way of causing a predator to give up the chase. It is a highly visible signal which says: 'I see you; therefore you have lost the element of surprise and there is no point in trying to catch me.' But, when the hyaena showed no sign of giving up, the buck broke into a run. Suddenly the night was alive with the jinking, flitting forms of the racing Tommies.

Twisting this way and that, the buck managed to stay ahead of the hyaena. At every turn he gained another precious yard, until in the end the big female gave up and slouched off towards Topi Plain.

There she was joined by two companions from the Murram Pit clan. After warily circling and sniffing each other in an elaborate ritual of greeting, they set off together towards the rim of the plain where the bunched wildebeest stood out darkly in the brilliant moonlight.

Wildebeest were the hyaenas' favourite prey, and their technique for hunting them was simple. There was no guile, no pretence at conceal- ment. They just charged the waiting wall of wildebeest, growling as they ran. When the animals stampeded, the hyaenas watched them carefully, looking for tell-tale signs of age or infirmity. Almost at once they spotted a certain awkwardness in the gait of a bull that marked him down as a target for slaughter. The bull pounded away in a straight line, instinctively galloping through another herd in an attempt to lose his pursuers, but they were not to be shaken off. Gone was the sloping, cringing demeanour they sometimes adopted when confronted by the Marsh lions. They ran with their heads up, rocking through the grass at an effortless pace which, if need be, they could maintain for miles.

As usual, the cub-killer was the first to reach the bull, snapping at his hocks long enough for one of her companions to grab him by the testicles. The moon shone on the whites of his eyes. In deep shock, he was hardly aware that they had opened his belly and begun to eat him alive where he stood. Eight minutes later his carcass was buried under a blood-soaked mob of pushing bodies as half the Murram Pit clan arrived to fight for their share of the kill. They ate hurriedly, giggling among themselves as limbs and bones were carried away to be crunched and swallowed out of sight of the pack. For more than an hour, the night was made hideous by the sounds of their feasting, and by the time they departed, nothing remained except the horned head and a dark stain on the earth.

The following evening, a sudden commotion in the Marsh told the hyaenas that the lions had caught a zebra. Before its squeals had died away, the big female and a dozen of her Murram Pit comrades had converged on the kill to find the three Marsh sisters crouching on the still quivering body. As the hyaenas stood in a cautious semi-circle, a fourth lioness emerged from the reeds, followed by two cubs. This was the Talek mother, whose twins were now old enough to accompany the pride on short hunting forays.

The lionesses began to feed, while the hyaenas keened in the shadows. The smell of blood emboldened them and they drew closer. The cub-killer felt the hair rising along the nape of her neck. With bristling tail held stiffly erect – an unmistakable signal of aggressive intent – she led the clan forward.

The lionesses were hungry and in no mood to give up their kill without a fight. As the hyaenas closed in, one of the Marsh sisters stood up and charged. The pack scattered into the night with a chorus of hysterical giggles, but moments later they came drifting back to encircle the snarling cats. As they closed in once more, the cubs tried to slink away, but the cub-killer darted in and bit one in the rump. The youngster yowled. Angrily the Talek mother whirled to defend her offspring. Her scything side-swipe sent the hyaena sprawling. Teeth crunched once, twice, in neck and flank, and the scavenger hobbled yelling into the reeds.

Meanwhile, the sounds of battle had attracted other hyaenas. Soon the Murram Pit clan was present in full strength, and the lions were forced to abandon their kill. Eagerly the hyaenas fell upon the stolen carcass, but they had paid a heavy price. When morning came, the sentinel marabous looked down from their trees and saw the lop-sided mask of the cub-killer grinning up at them with lifeless eyes.

*   *   *

. . . the lop-sided mask of the cub-killer grinning up at them with lifeless eyes . . .

They would file out of the forest to bathe and feed in its reedy deeps . . .

All through June and July the Marsh had remained so wet that only the elephants could reach the heart of it. Led by the ancient one-tusked matriarch they would file out of the forest in mid-afternoon to bathe and feed in its reedy deeps. But, as the surface water evaporated during the dry season, the marsh became increasingly accessible. Now the musty taint of the big, shaggy-coated waterbucks hung in the reeds; and every day, when shadows were shortest, buffalo lumbered out of the forest to drink and wallow in the mire. Soon even the Marsh lions themselves were able to reach its innermost recesses, and the shy reedbucks resting among the sedgy tussocks began to add their shrill whistles of alarm to the multitude of marshland voices.

By the time the first of the migrating herds had begun to arrive, the buffalo had already made deep inroads into the shrinking Marsh, and as the daily tides of zebra and wildebeest grazed and trampled their way to water, the reeds retreated still further. Yet even in the driest months there always remained a dense tract of sedge, a species akin to papyrus, but with curious articulated stems; and it was here that the Marsh lions gathered as the migration reached its peak.

The reeds were their sanctuary and their abattoir. Here was everything they needed: cool shade, permanent water and perfect concealment from their prey. As pools and water-courses dried out on the open plains, the wildebeest became increasingly dependent upon the marshland spring and its muddy runnels. Every day they would move into the Marsh, and every day the lions would kill them. Sometimes, arriving early in the morning, the herds would see the lions disappearing into the sedge to escape the heat of the sun, and would gallop away with a volley of snorts. But it made no difference. Sooner or later other, unknowing wildebeest would arrive. Then, as the thirst-crazed beasts spilled into the Marsh, the reeds would erupt as the familiar tawny shapes emerged to claim yet another victim.

The migration was the season of plenty, and the lions grew fat on the profligate herds. Their coats acquired a lustrous sheen. Their bellies hung heavy with meat. So easy was the hunting that even the ponderous and shaggy-headed Marsh males were able to kill occasionally for themselves rather than take their share of the animals pulled down by the lionesses.

. . . shy reedbucks resting among the sedgy tussocks . . .

Even the ponderous and shaggy-headed Marsh males were able to kill occasionally for themselves . . .

Day or night, there was no respite. Flying in panic from an ambush in the Marsh, wildebeest and zebra would gallop away through the swaying grass only to run headlong into the fatal embrace of Brando, Scar or Mkubwa.

Although the males still patrolled the perimeters of their territory, their lionesses seldom strayed far from the Marsh and its immediate surroundings. Relieved of the need to make long hunting forays across the plains, they were content to lie up in the reeds and pick off the wildebeest whenever they were hungry. Bloated with flesh, they became finicky in their eating habits, taking only the choicest morsels. If they killed a cow buffalo, they would merely chew at it, eating the loins, ears and soft facial tissue, and leave the rest for the scavengers. Their abandoned kills littered

Its red glow flooded the Marsh with lambent light . . .

the grasslands around the Marsh, and were marked each morning by fresh spirals of descending vultures.

Sometimes, when the wildebeest practically stumbled over them in their eagerness to drink, they killed even when the stimulus of hunger was absent. It was not killing for pleasure; only the innate response of a predator confronted by an opportunity too easy to ignore.

For the Talek mother in particular, the closeness of the wildebeest seldom failed to arouse her hunting instinct. She lay now quiescent in the reeds, her breath steaming in the damp dawn air. The sun rose. Its red glow flooded the Marsh with lambent light. The tall stems striped her flanks with tigerish shadows that made her almost invisible. She seemed oblivious to the presence of her twin cubs as they cuffed at each other in

They were spilling down through the yellow grass towards her . . .

mock combat, though she growled softly when their play became too boisterous.

Yet even at rest, her senses were unconsciously unravelling and interpreting the scents and sounds of her marshland world. Her big round ears missed nothing. She heard the harsh croak of sacred ibis, the yelp of a fish eagle, the scream of elephants in the forest. Her nostrils drew in a tangle of ripe odours: geese and reedbucks, buffalo dung, rotting flesh, lion, grass, wildebeest, and the black ineluctable reek of the Marsh itself.

With eyes half closed against the light, her face wore an expression that could have been mistaken for serenity. But her dreamlike repose was a mask. Heat, thirst, hunger, pain, danger – these were the limits of her world. Of the emotions known to human beings she felt only fear, anger, distress and the sense of well-being that came with a full belly. Her gaze appeared to be fixed on some remote horizon of the mind, but what she saw were the movements of wildebeest out on the plains. From the edge of the reed-beds, she watched the dust rising as the herds began to converge once more on the Marsh.

As they drew nearer she could hear their sonorous grunts. By mid-morning they were massed along the ridge above the spring, and there they faltered, snorting uneasily as their nostrils caught the taint of lion. From her hiding place in the reeds, the lioness could see the glassy glitter of their horns sharp-etched against the sky. Then they were spilling down through the yellow grass towards her.

At once her whole demeanour changed. Her jaws parted slightly, giving her a look of eager alertness. Her ears cocked forward, and a fierce light filled her widening eyes. Sensing their mother's change of mood, the cubs ceased their sparring and crouched beside her.

As she watched, she heard the soft swish of a body pressing through the Marsh behind her. The reeds parted and the head of another lioness emerged. It was one of the Marsh sisters. She, too, had heard the wildebeest and had stolen forward to join her pride companion. By now the first wildebeest were drinking from the spring in the company of a large herd of zebra. Soon the pool was completely surrounded, forcing the latecomers to spread out along the stream which meandered into the reeds. Slowly, as the multitude expanded, the animals on the edge of the herd were forced closer to the waiting lions.

The nearest animals were now no more than thirty paces away. Among them was a zebra stallion and his three mares. Cautiously the stallion lowered his head to drink. A sudden breeze drifted across the Marsh, shaking the reeds with a dry whisper. The stallion looked up, and found himself staring incredulously into the eyes of the Talek lioness.

Zebra and wildebeest stampeded together, cannonading into each other in their desperation to avoid her charge. As they fled across the stream, the second lioness launched her attack, throwing them into even greater confusion. Their hooves hammered on the dry ground as wave after wave reached the safety of the open plain, but for one floundering wildebeest there was no escape. Panicked by the appearance of the second lioness, she had blundered into a wallow and sunk up to her belly.

The wildebeest did not struggle when the lionesses pulled her from the mud. Exhausted by her efforts to free herself, she kicked feebly, then lay

The wildebeest did not struggle when the lionesses pulled her from the mud . . .

inert with heaving flanks where they had dragged her. The cats lay down beside her, but made no attempt to strangle her. Their bellies were still full, and they quickly lost interest once the victim was caught.

Not so the cubs. At first they were afraid of the strange creature with the curling horns, and peered at it warily from behind their mother. But when a third lioness arrived and began to open it up in front of them, their confidence grew. Soon they were trying to join in. Unsure of what to do, they jumped on the defenceless body, fell over its neck and pounced on the twitching ears. Unable to rise with the lioness across her withers, the wildebeest could only lift her head in protest. Instinctively the bolder of the two cubs sprang forward and wrapped its paws around the long muzzle. Her head rolled glassy-eyed as the cub's sharp teeth sank into her nose, trying to imitate the classic bite which lions often use to kill their victims, and although the cub was still far too small to deliver the *coup de grâce*, the incident showed how even at such an early age the young Marsh twins were acquiring the hunting skills which would sustain them in later life.

6

October-December 1978

The Departure of the Herds

By October the herds had laid the glasslands bare. In their endless foragings, following the showers across the plains, they had crossed and recrossed the luggas and ridges until only thornbush and the bitter yellow Sodom apples remained. All green had long since gone, burned to a dusty harvest gold. Where the deep oat-grass prairies had run rippling to the horizon, nothing was left but chaff and stubble.

For weeks an implacable drought had gripped the Mara. So little rain had fallen that the Maasai drove their cattle deep into the reserve to water them. At night the escarpment glowed with angry necklaces of flame where they had fired the grass to hasten new growth. At midday the heat lay like a mist on the plains. Animals fell sick. Every morning the Ruppell's griffon sailed out to find fresh corpses. Slumped in the remaining reed-beds, the Marsh lions groaned in hot, restless sleep. Flies tormented them. Tempers shortened, and the three pride males took themselves off to the forest where they could rest undisturbed. The land seemed drained of life. Giraffes and impala sought the shade of trees. Only in the tawny sunlight of early evening did movement return, when the cheetahs awoke and gazelles bounced away over the bare earth.

Now the wildebeest felt the oncoming season of the short rains. Away to the south the first showers had already burst. Lightning forked along the dark horizon, and the herds turned towards it, anxious to return to their calving grounds. That night, the first of the zebra departed for the Serengeti. Their hooves clicked as they picked their way over the stony ridges to Paradise Plain and headed for the river-crossings. The wildebeest followed, and soon the hunting grounds of the Marsh pride lay deserted. Of the herds which had blackened Topi Plain and poured into the Marsh each day to drink, only a scattering of old wildebeest remained. Too sick to make the return journey to the land of the Gol Kopjes, they stood in the silence of the empty grasslands, waiting for their lives to come full circle.

All over the Mara the migrating herds were leaving the reserve, streaming away towards the Tanzanian border. Already the first flying columns of zebras had crossed the Great Sand River and were back in the northern woodlands of the Serengeti. Along the rocky bed of the Talek, only a string of stagnant pools remained where giant fig trees cast their protective shade, and the animals which had been wandering between the Ntiakitiak and Lolongabulu rivers crossed it with ease as they poured down towards Keekorok across the Burrungat Plain. Only the Mara River still flowed strongly. Although there were now waist-deep shallows where the herds might wade across in the dry season, it remained a formidable barrier for the zebra and wildebeest which were now massing along its banks at the southern end of Paradise Plain.

When the migration had first arrived in the Mara, the river had still been high and swollen after the rains. Now it had fallen dramatically, exposing scattered rocks that lay like smooth stone hippos in midstream. Here the water ran no more than rib-deep and, when the time came, this was where the herds would cross, stumbling over the bones of animals which had perished during the crossings three months earlier.

Hidden in thick cover beside the river, a lioness and four large cubs from the Paradise pride were woken by the noise of approaching wildebeest. They had been there since morning, but the converging horde never knew they were being watched. During the night, the rest of the pride had gone hunting on the plain below Rhino Ridge. There, joined by the Paradise males, they had killed a zebra and were now lying up in the rocks, but the lioness, having chosen to hunt alone, had gone hungry. At dawn she had surprised a group of zebra drinking at the water's edge, but they had stampeded across the river towards two more lionesses who watched from the other side.

The strangers were members of the Serena pride which frequented the area around Ol Doinyo Oseyia, the high ridge on which stood the Mara Serena Lodge. Although their territory lay across the river in the Mara Triangle, the water was no obstacle to the Serena lions. Even when it was too deep to wade they could swim it with ease to trespass in the hunting grounds of their neighbours, the Paradise pride. Drawn by the zebra's urgent barking, they had come down to the river to investigate. When they saw what was happening, one lioness sank into the grass while her companion began a long stalk. Slowly she crept closer. For the last few yards she was completely in the open, yet the zebra did not see her until she charged.

. . . a group of zebra at the water's edge . . .

The last animal was a young mare . . .

Her paws threw up torrents of spray as she lunged in their wake . . .

Her paws threw up torrents of spray as she lunged in their wake, stampeding them back across the river, but the current slowed her down. One by one they floundered ashore and galloped away across Paradise Plain. The last animal was a young mare, not quite fully grown. As she clambered over the crest of the bank the lone Paradise lioness reared out of the ground and pulled her down.

The zebra died quickly, caught by the windpipe. Hurriedly the lioness opened her and began to feed, but she had not taken more than a few mouthfuls before she was disturbed by the approach of the two Serena lionesses from across the river. Outnumbered by the intruders, she was forced to let them settle down to share her kill, and for a few moments they fed amicably together, as if all three belonged to the same pride. But the truce could not last, and the two Serena lionesses finally rounded on her and drove her away.

When she had gone, one of the Serena pair stood up and walked back across the river, pausing to drink under the shade of green palms. Her companion continued to feed for a while but was clearly ill at ease in the territory of the Paradise pride. In the end she, too, stood up and dragged the half-eaten zebra to the edge of the bank, whence it toppled into the water. Then, straddling the heavy carcass with her front legs, she grabbed it in her jaws and began to carry it across the river. Her neck and shoulder

muscles bunched with the effort, forcing her to pause for breath at every few steps, and the slippery rocks made it hard for her to keep her balance in the current. Once she stumbled and almost went under, but did not release her grip until she reached the Serena bank. There, panting with exhaustion, she hauled the carcass from the water and withdrew into the bushes to rest.

In the meantime, the Paradise lioness, having lost her kill, had wandered off along the river to join the four cubs. Seeing the dust clouds raised by the approaching wildebeest, she had led them towards the crossing-place and had crawled into a dense patch of bankside cover. Awakening now in late afternoon, they felt the ground tremble as the wildebeest pressed forward. Above the chorus of grunts they could hear the splash of falling bodies as the first animals plunged into the water, followed by the bleat of calves and the bellows of the injured. Then for a long time there was nothing but the rumble of the herds as they fought their way across the river and thundered off towards Ol Doinyo Oseyia.

The two Serena lionesses finally rounded on her and drove her away . . .

She stood up and dragged the half-eaten zebra to the edge of the bank . . .

The slippery rocks made it hard for her to keep her balance in the current . . .

By early evening the wildebeest were gone. The last groups of stragglers had crossed the river and disappeared over the horizon. But then, as the sunset glowed eerily through the dust, a solitary calf came hobbling out of the emptiness of the plain. Stampeded by hyaenas, it had snapped its leg in a pig-hole and had fallen far behind. Now its wanderings were almost over. The river ran like molten gold in the dying light. The grasshoppers fell silent as dusk engulfed the cooling plains. Slowly, on its three sound legs, the calf came faltering forward, unaware of the cats in the shadows. The last thing it saw was the face of the Paradise lioness as her jaws gripped it by the throat.

Towards midnight, when lioness and cubs had eaten their fill, they ambled off across Paradise Plain. There they met three more of the Paradise lionesses, accompanied by the biggest of the pride males, and stayed with them for the rest of the night. Morning found the whole group sleeping among a scattering of old termite mounds; but, as the sun rose higher, the heat forced them from the plain. Irritably they arose and stretched, walked a few paces and slumped again, as if too tired to move. But eventually the lioness with the four cubs walked off towards the river, and the rest of the pride followed, with the big male at the rear.

Alert on their ant-hill watch-towers, statuesque topi watched the lions file past, but the cats ignored the snorting antelopes and continued on their way to the river. One by one the lions disappeared into the dim green tunnel of a hippo trail, and emerged at the other end in bright sunlight where it opened to the water's edge.

Palm trees arched across the river, their graceful fronds hanging limply in the breathless air. Beneath them, the floodwaters had gouged a wide pool where the river swung in a sharp bend against a high bank of sun-baked clay, and here a bull hippo had made his home. The hippo was permanently disfigured by an impacted upper tooth which had grown back through his upper lip and burst through the thick whiskery hide, where it showed like an ulcerous third nostril. The pain no longer maddened him and the flesh had long since healed, but it had left him irritable, morose and unpredictable. His needs were simple. At night he left the river by the same well-worn trail to graze under the stars. By day he lazed in the water, yawning and snorting, occasionally whirring his tail to spread a fine stream of manure through the buoyant brown murk of his pool. He lay now, enjoying the sensation of soft nibbling lips as a host of small, carp-like fish cleansed his hide, grazing on the algae which bloomed on his ponderous flanks. Normally the pool was deep enough to allow him to submerge completely if danger threatened, but the dry season had so reduced the river that the water would barely cover him.

When the lions appeared on the bank above him, the hippo gave an ill-tempered snort. His nostrils forced a fine spray of water into the air as he tried to sink out of sight. Yet even with his short legs doubled under him and his belly touching the muddy bottom, the sun still shone on his back. His vulnerability nagged at him, causing his absurdly small ears to twitch angrily, flicking away beadlets of water.

Palm trees arched across the river, their graceful fronds hanging limply in the breathless air . . .

The lion cubs were eager to drink but could sense their mother's caution and kept well behind her as she walked towards the water. Halfway down the bank she stopped and stared back at the pair of frog-like eyes watching her from midstream. Her tail lashed from side to side as she hesitated, sensing how dangerous a cornered hippo could be. When he did not move, she padded forward again with the cubs tumbling after her, and lowered her head to lap at the water's edge.

This was more than the hippo could endure. Water swirled off him as he lifted his huge angry bulk and shook his head. Then, with a cavernous yawn of rage, he turned and faced her. The startled lioness took one look at the gaping pink maw, the wickedly jutting ivory teeth, and fled. The hippo gave a triumphant snort and sank back into the pool, but no sooner had the ripples subsided than a new problem arrived.

The big pride male had lagged behind his females. Now he was hot and thirsty. He moved slowly, his great heavy head lolling, dry-tongued and panting. One of the lionesses rubbed her cheek against his mane, but he ignored her attentions and padded imperiously down to the river; the pride turned to follow.

For a moment the hippo seemed to shrink back into the pool. Unable to submerge, his tiny brain was wrestling with the possibilities of escape. Behind him, the bank rose in a twenty-foot vertical wall. Upstream, he

. . . the gaping pink maw, the wickedly jutting ivory teeth . . .

Despite his enormous balloon-like girth, he moved with the speed of a charging rhino . . .

knew to his cost, was another pool with many hippos, and his presence there would almost certainly provoke an attack. Downstream, the river ran noisily over rocky shallows. There was only one way out.

He gave no warning. The lion saw the pool erupt, saw the water part in two crashing bow waves as the hippo ploughed straight for him. He came with head down and, despite his enormous balloon-like girth, he moved with the speed of a charging rhino. He pounded up the bank, blunt mouth slightly parted to reveal a gleam of ivory, scattering the Paradise pride in all directions. But he did not pursue them far. At the edge of the bushes he stopped, defecated with furiously wagging tail, then turned and trotted back to his pool. As the cool waters closed around him, he felt again the soft attentions of the fishes. The lions faded from his memory as his nostrils flared and a deep, mirthless guffaw shook from his throat.

*   *   *

With the departure of the migration the Mara seemed strangely empty. There was still plenty of game in the reserve. Some zebra remained all year round. Giraffes stood out on the horizons, moving with their slow, dreamlike gait. There were reedbuck and waterbuck, eland, kongoni and large numbers of buffalo, warthogs, topi and gazelles. But without the drama and mystery of the great herds, the plains no longer teemed with life. Cropped by the millions of nibbling mouths, the grasslands lay dormant. In the still afternoons, flat-bottomed clouds sailed across the sky, trailing islands of shadow over the heat-hazy savanna. Soon it would be the time the Maasai called *ilkisirat*: the month of the short rains, when showers would bring a brief respite and grass would spring in fresh green flushes wherever the erratic scuds of rain washed the land.

For the Marsh lions, and for all the Mara prides, the days of easy hunting were over. Bloated with meat and offal, they had lived royally for as long as the wildebeest had darkened the plains. Now the herds had gone they would have to work harder for their kills, foraging farther from the Marsh to hunt on the open plains. Some of the other predators had moved south with the migrating herds. The wild dogs of Aitong had not been seen since Scar chased them off a kill in September, and they were almost certainly on the heels of the wildebeest. Most of the nomadic lions had gone, too, the outcast veterans and sub-adults; but the prides remained, and the presence of the territorial males established a sense of order, regulating the numbers of hunters the land could sustain through the lean times.

Other hunters also remained. For the cheetahs, the disappearance of the wildebeest made little difference. Although one kill in ten might be a wildebeest while the herds were in the Mara, the spotted cats continued to feed almost exclusively on Thomson's gazelles, impala and topi fawns. The reserve was ideal cheetah country. Prey was abundant, and the terrain provided both the open spaces in which to outrun the fleet gazelles and sufficient cover in which to rest or hide from other predators. Slowly, over the years, in spite of poaching and harassment by lions, hyaenas and Maasai dogs, their numbers had increased, and there were now perhaps a hundred cheetahs in the Mara region.

Unlike the social carnivores – lions, dogs, hyaenas – the cheetah is essentially a solo hunter. It lacks the strength of the lion, the ferocity of the leopard, the crunching bite of the bold hyaena. Unlike other cats it cannot fully retract its claws, which are blunt and dog-like with constant running. Its jaws are small. It is a large animal, but not strong enough to defend its kills or its cubs against the determined predators whose territories it shares. Only its constant alertness and its ability to hide and run enables it to live in uneasy co-existence, exploiting the same hunting grounds as its more powerful neighbours.

Above all, it relies on speed. The rakish, loose-limbed cheetah with its sway-backed racing frame is the culmination of an ancient and inseparable bond between hunter and hunted. It has evolved stride by stride with the gazelles, the slenderness and quicksilver pace of the one being matched by the feline grace and devastating acceleration of the other. The small round head, the deep chest and the elastic backbone slung like a hammock between its haunches, everything about it is designed for the short fast rush. In the precarious world of the cheetah, every day is a race for life. Without its speed it would simply starve. That is why it has become the fastest animal on earth, said to be capable of reaching seventy miles an hour.

. . . the small round head, the deep chest and the elastic backbone slung like a hammock . . .

For Nefertari, resting in a thicket on the edge of Topi Plain, survival was even more of a struggle. No other cheetah in the Mara had to cope with the travails which she endured. Like all her kind, she did not have a fixed territory, since the term implies a defended area, clearly an impossibility for a lone cheetah. But she did move within a regular home range of five or six square miles, and it was her misfortune that it was overlapped by the territories of both the Marsh and Miti Mbili lion prides, and overrun by the Murram Pit hyaenas. Furthermore, her proximity to Governor's Camp meant that she was persistently watched and followed by convoys of tourists. She had long since lost all fear of vehicles, knowing they meant her no harm, but sometimes they inadvertently interrupted her hunting, forcing her to go hungry or to hunt by moonlight, when other predators were more likely to steal her kills.

Now she was even more vulnerable. Twelve weeks earlier she had mated with the big male known as Badfoot, who had limped with a swollen right leg for most of 1978 but had somehow managed to stay alive and recover. Afterwards, she and the male had gone their separate ways, each cat fending for itself, but in the meantime her advanced pregnancy had begun to affect her hunting. She was like a racing machine, so finely balanced that even the few extra pounds of unborn kittens could tip the scales against her. Now, when she hunted, she missed more often than she killed, and her days became a treadmill of long fruitless searches across the arid plains.

Nefertari needed the short rains just as much as the grazing animals did. The Thomson's gazelles which were her favourite prey had no fixed breeding season. They gave birth throughout the year, but they seemed to favour the green places where showers had fallen. If the rains came soon, Nefertari would eat well. At first, the fawns could not run fast. They hid in the grass, lying with their chins on the ground to make themselves as inconspicuous as possible. They had no scent; their mothers licked them clean. But Nefertari was experienced in the ways of gazelles and would have little difficulty in finding them.

Nefertari yawned again . . .

She lay now at full length on the ground, swollen with her growing cubs. A veteran of many seasons in the Mara, her pale coat had lost its once youthful lustre, though the markings on the end of her tail remained her most remarkable feature. Instead of the usual black and white rings it was marbled with curious black hour-glass spots. Her grey muzzle and the grey hairs which blurred the black spots on her nape and crown gave her a grizzled look which befitted her age. With her characteristic black tear-stains and sad, enigmatic face, she was in many ways a tragic figure. Of the five litters she was known to have produced, three had been lost to predators, and her last litter drowned by a flash flood. Only once had her cubs survived to adulthood: two sons and a daughter so tame that they would even climb on to the roof of the Land-Rovers from the Mara River safari camp. There was something profoundly moving about her stoicism. She was the oldest female cheetah in the Mara but she still carried new life inside her.

Every day for the past two weeks, clouds had built up through the sultry afternoons. Long before dusk the escarpment lay hooded in thunderheads. Rain threatened, but did not come. The clouds cooled the air and the cheetah stirred. She yawned and began to lick the fur of her chest, but remained on her side to hide from unfriendly eyes. Caution and concealment were second nature to her, and even when she raised her head briefly to peer through the fleck and speckle of the tufted grass-heads at the plains beyond, she did not change her position.

Somewhere nearby, red-necked spurfowl began to call. Briefly the sun reappeared between the cloud base and the horizon, flooding the long swell of the open plains in a brilliant amber light. Nefertari yawned again, then rolled on to her stomach, resting with hind legs drawn in beneath her, and head raised. It was two hours before nightfall, the favourite hunting time for cheetahs, and as she stared, sphinx-like, across the glowing ridges, her agate eyes burned more fiercely as they fixed upon the distant shapes of gazelle.

Satisfied that there were no enemies hidden nearby, she got up and began to walk slowly towards the herds. On the way she paused beneath a solitary wild olive and spent several minutes sniffing intently around its base. The limping male, Badfoot, had passed by recently, and had marked the trunk with his scent. Anxiously she scanned the plains in case

. . . the Tommies – dainty brown animals with black and white rumps and elegant leaf-shaped ears . . .

he was still within range. Had he been there, she would have gone out of her way to avoid him, for she would tolerate male cheetahs only when she was in season. But once she had reassured herself that he had gone, she resumed her steady progress towards the gazelles.

The Tommies are the most prolific of the smaller herbivores: dainty brown animals with black and white rumps and elegant leaf-shaped ears. Males and females alike carry horns and sport a distinctive dark horizontal stripe along the flanks above their white bellies. To a casual eye they would appear to be scattered across the plain at random but, as always, the impression of disorder is misleading. What seems to be an aimless confusion of wandering animals is in fact a precise and complex social order, the separate herds of females and bachelor males revolving around the fixed territories of the adult breeding bucks.

As the cheetah approached, the herds parted to let her through. They did not run, but stood watching nervously, stamping and snorting as she walked among them. After she had passed, they closed ranks behind her and followed at a cautious distance, as if mesmerized by the presence of the killer in their midst.

It was a good time to hunt. Often when she hunted in the mornings, the falling shapes of vultures, bateleurs and tawny eagles would alert other predators to her presence. Many were the kills she had lost to lions and hyaenas in this fashion. But now the sky was empty. Her pace quickened. She began to canter, scattering the gazelles like leaves in the wind. The canter became a gallop and the gallop a sprint as she closed on her target. She covered the ground in enormous bounds, her hind legs stretching in front of her forepaws as she drove forward on the heels of the scudding gazelle. But, just as she was about to deliver the underhand swipe that would crumple the prey in a tumble of flying hooves, her stamina gave out and she halted in a flurry of dust.

It was several minutes before she had recovered sufficiently to resume her search. The burden of pregnancy had reached a critical point. Her body cried out for fresh meat, and only an hour of daylight remained. Once again she walked towards the gazelles, and again they followed her progress across the plain; but this time she ignored them and climbed an ant-hill. Giant cliff-heads of raincloud towered along the eastern horizon; against them, etched in sunlight, the golden cat stood out against the gathering storm. From the top of the mound her gaze swept the grasslands. She saw zebra, giraffes, a column of topi cantering away to Miti Mbili, and at last her sharp eyes found what she had been looking for.

The lone gazelle lay quietly beyond the edge of the herd, in the very act of giving birth. The contractions were coming strongly now, and the forelegs of her baby appeared quickly. When the head emerged, she stood up. The last rays of the sun glistened on the foetal sac as the tiny bundle dropped to the ground. Anxiously the mother looked round, then began to lick her newborn infant as it struggled and squirmed from the wet membranes.

The gazelle saw the cheetah just as the fawn was staggering to its feet, and at once she pranced off, hoping to divert the predator's attention while the fawn instinctively fell back into the grass, to freeze in the chin-to-ground posture which was its only means of avoiding detection. For a moment the cheetah was confused. When the mother had fled, Nefertari had taken her eyes off the crouching fawn, but her searching gaze picked it out. She approached the uncomprehending creature and dabbed at it with her forepaw, as a tabby might tap at a ball of wool. The fawn wobbled to its feet and tried to run, but could manage no more than a dozen steps before it was knocked down and seized by the throat. The mother gazelle watched helplessly as the cheetah carried the feebly struggling form away to the edge of a thicket and lay down to feed. Ten minutes later, as the first spots of rain began to fall, there was nothing left.

In the second week of the short rains, Nefertari gave birth. During the night she crawled under a thorn tree near the edge of Leopard Lugga and produced five cubs. She had chosen her lair with care. Up-ended by a hungry elephant, the tree had fallen on its side but had continued to grow. Where its canopy touched the ground, lank grasses had shot up, entwining themselves among the tangled branches to form an awning as thick as a hedge behind which the cubs, mere scraps of wet black fur, were well concealed. Unlike the infant wildebeest and gazelles, which could run almost as soon as they were born, the cheetah cubs were utterly dependent on their mother.

Next day she left them to go hunting on Topi Plain. The cubs slept, almost invisible in the shadows of the thorn bush. When she returned Nefertari did not go directly to the lair. A slight movement in the grass at the edge of the lugga had caught her eye. For a long time she stood poised, her tail held out in a taut curve as she scanned the bank. A moment later the grass moved again, and she saw that another cheetah was watching her. It was the male, Badfoot. Nefertari crouched. She dared not return to her cubs because the male would almost certainly kill them if he found

. . . the chin-to-ground posture which was its only means of avoiding detection . . .

The cheetah carried the feebly struggling form away . . .

She moved them again, dropping them in a patch of long grass . . .

. . . the forlorn figure of Nefertari . . .

them. Anxiously she watched him prowl along the bank, nosing among the bushes as if deliberately searching for the cubs. Yet even though he passed within a dozen paces of their hiding place, he did not see them. Eventually he emerged from the lugga and padded aggressively towards her, but she hissed so furiously at him that he turned aside and eventually wandered off across the plain.

Nefertari waited until he had gone, then returned to the lair, greeting her family with a strange, bird-like chirrup as she lay down and allowed them to suckle. But the incident had unsettled her, and afterwards, when the cubs were sleepy with milk, she picked them up and carried them one by one to a new hiding place.

When the cubs were thirteen days old, she moved them again, dropping them in a patch of long grass not far from the spot where one of the game-viewing tracks crossed the lugga. She lay beside them all morning, sleeping and suckling; but during the afternoon, as the time for hunting approached, she moved to a nearby termite mound. Like all cheetahs, she regularly used the ant-hills as vantage points from which to look for prey. Now, as she scanned the plain, her body stiffened and she uttered the low moan that cheetahs use when danger threatens. Advancing through the grass in a menacing wedge was the entire Miti Mbili pride, led by the unmistakable ragged-eared figure of Notch.

The sudden appearance of the lions took Nefertari completely by surprise. The six cubs accompanying the pride were now half-grown and, when they saw the cheetah silhouetted on the ant-hill, two of them broke away from the close-knit phalanx and charged. Nefertari fled, but stopped and turned when she realized she was not being followed.

Notch, meanwhile, had found and killed three of the cheetah cubs in the grass at the base of the ant-hill. She crushed them in her yellow teeth and cast them aside where they were later found by the Murram Pit hyaenas. But before she could kill the last two kittens a Land-Rover arrived and drove her off. The driver took the cubs to Governor's Camp where a heated debate followed. Should they be flown to Nairobi, to spend the rest of their lives in captivity? Or should they be returned to take their chance in the wild? In the end, the decision was swayed by a visiting vet, who suggested that an attempt should be made to try and return them to their mother.

Twilight was not far off. The day had been overcast; now it had begun to drizzle. In the failing light, the driver managed to pick out the forlorn figure of Nefertari. She did not run when the vehicle drew up beside her. Gently the driver placed the cubs in the grass. Afterwards, the watchers heard the cheetah chirrup as she greeted her babies. Tenderly she picked up each one in turn and carried it away to the fig trees beside the Marsh.

The two cubs were never seen again. Next morning, when the driver returned to look for them, a lioness and her two large cubs arose from the grass where Nefertari had last been seen with her offspring. It was the Talek mother with the young Marsh twins.

7

January-May 1979

The Miti Mbili Pride

FOR A WHILE, the loss of Old Girl weakened still further the precarious tenure of the Miti Mbili lions. Surrounded by vigorous and hostile prides, and still lacking the protection of regular adult pride males, they missed the reassuring presence of the grizzled old lioness. Without her the group had lost its old cohesion. Young Girl would sometimes hunt apart from her companions, taking with her the three cubs sired by Brando. The cubs were now nearly eight months old, and big enough to accompany their mother. Notch was now the oldest of the lionesses. The seven cubs born to her and Shadow were robust two-year-olds, able to hunt and kill, albeit clumsily. They still relied on the adult lioness to subdue zebra and other large prey; but as they entered sub-adulthood their growing strength added greatly to the power of the pride. They were now about two-thirds grown, and already the four young males were beginning to develop the broad heads, strong shoulders and the first wisps of the heavy manes they had inherited from their fathers.

Like the Marsh lions, Notch and her family had lived well for as long as the migrating herds had roamed the Mara. But while the Marsh pride had usually ambushed the wildebeest in daylight as they came to drink, the Miti Mbili lions had hunted mainly at night, stalking along ridge and lugga when the herds returned to the open plains. After the wildebeest had

The cubs were now nearly eight months old, and big enough to accompany their mother . . .

Notch was now the oldest of the lionesses . . .

left in October, the lions had fallen on the luckless warthogs, whose young had just begun to venture from their burrows. Topi were also plentiful in the short rains, and for a time the Miti Mbili pride grew fat again as the wandering herbivores followed the showers across their hunting grounds. But the grass rains never lasted long. Soon the hot, clear days of December and January had withered the new growth, and the plains lay grey and ashen as if the rains had never been. Game became scarce as first Miti Mbili Plain and then Topi Plain emptied in the general drift towards the river, and the prides were gradually drawn westward in the wake of their prey. Under cover of darkness, Notch and Shadow began to lead the young lions ever deeper into their old domain. And when they found no sign of their enemies on the eastern side of the Marsh (for the Marsh lions meanwhile had taken to hunting between the reed-beds and the river), they became bolder still. Instead of retreating during the day to lie up in the safety of their own home range, they took to resting under an isolated tuft of tall trees that grew between the lower end of Leopard Lugga not far from the spring.

Not long after the killing of Nefertari's cubs, the two adult Miti Mbili lionesses and their cubs were encamped on the lower reaches of Leopard Lugga. During the night they had caught and quarrelled over a topi. Normally they fed with little enmity, but times of scarcity always sharpened their uncertain tempers and on this occasion their squabblings reached such a pitch that they were heard by the Murram Pit hyaena clan. They remained in the lugga all morning, protecting what little was left of the kill. The sun burned fiercely through the thickets, mapping their bodies with sharp thorny shadows. Flies tormented them, swarming in the white fur of their upturned bellies as they twitched and flinched in growing annoyance. Tongues lolling, flanks heaving, they lay scattered along the edge of the dried-up water-course, with Notch lying apart from the rest of the group.

Motionless upon a termite mound, a solemn topi watched them pass by . . .

By noon the sun was unbearable and the lions were forced to move from the lugga into the deeper shade of the trees. Notch led the way, with Shadow and the six large cubs following at a respectful distance. They moved sluggishly, padding heavy-footed through the crackling heat, as if their limbs were still drugged with sleep. Motionless upon a termite mound, a solemn topi watched them pass by: eight tawny cats in the tawny grass. Soon, all the topi could see were the black tufts of their tails and the dark markings behind their ears, so perfectly did their coats blend with the sun-dried herbage.

What Notch could not know was that earlier that day the Marsh lions had moved from the nursery forest at the southern end of the reed-beds to drink at the spring, and were now resting under the same clump of trees which the Miti Mbili pride was approaching. Fortunately for Notch and her family the three big Marsh males were not present – they were lying up in the forest near Governor's Camp – but all five Marsh lionesses were there, together with the young Marsh twins and the five small cubs.

The Marsh lions had moved from the nursery forest to drink at the spring . . .

If Notch had known what lay beyond the rise she would almost certainly have turned and slunk away. Despite her cantankerous nature, she had no wish to provoke a confrontation, least of all as a trespasser in alien territory; but the rising ground meant that neither group was aware of the other until it was too late. The young Marsh twins reacted first, taking fright and fleeing into the scrub; and the smaller cubs, sensing the sudden change in their mothers, crouched in the grass. But the five adults rose as one and advanced to meet the intruders.

They came with heads low, as if stalking prey. One of the Marsh sisters began to roar; a short, harsh roar, full of violence. Although they were outnumbered by the Miti Mbili lions, this was the very heart of the Marsh lion territory and they were not about to give way. When they were about thirty paces away they broke into a run and were met head on by Notch and Shadow.

The battle was short and vicious. The two prides collided, grappled and broke away, snarling. Already blood was flowing. Almost im-

They came with heads low, as if stalking prey . . .

mediately they clashed again, flattening the grass as their bodies locked and rolled, rising in a flurry of swiping paws and falling back again in writhing mêlées. In the thick of it was Notch, bleeding profusely from lacerated flanks, but fighting furiously. She heard the gasps of her opponents as her canines and claws clenched on fur and flesh. All the Marsh lions were fighting now. Two of them had isolated one of the young Miti Mbili two-year-olds and would surely have killed him if Shadow had not come to his rescue. As the youngster fled, the will of the pride seemed to crumble. They turned and ran, pursued by the victorious Marsh lions, and did not pause until they had disappeared from view beyond Leopard Lugga.

The Marsh lions roared, reasserting their right to the heartland of the pride, and when no answering roar was heard they returned to the shade to lick their wounds. Meanwhile, Notch and Shadow were leading their family back towards Miti Mbili. They had been badly mauled, and from that day on, although they still made occasional incursions into the Marsh lions' territory when game was scarce, they were never again seen near the Marsh.

*   *   *

Once again the fortunes of the Miti Mbili lions had reached a low ebb. Driven from their old dry-season hunting grounds, they were forced to make long journeys each night across the plains. With the grass gone, there was little cover except along the luggas, and what little game remained there had become wary, having learned to avoid the lions' favourite ambush places. With only two fully grown lionesses and no males to assist them the pride found buffalo hard to kill, and, had it not been for the ubiquitous warthogs, some of the young lions might have starved. Even then, the presence of the over-eager and inexperienced youngsters often botched a hunt, alerting their intended victims before the cats had stalked close enough to make a successful rush.

The pigs kept them alive, but only just. Their bodies grew thin. Their wounds itched. Gaunt-ribbed and fly-speckled, they lay panting through the long hot dusty days, scarcely moving except to stand up and flop down again whenever the creeping shadows of their tree-shelters left them stranded in the sun. Nightfall released them across the plains again, endlessly prowling in search of food. But although they killed regularly, there was never enough to go round.

The lone huntress, Young Girl, was also finding it increasingly hard to feed herself and her cubs. Padding through the starlight with the youngsters at her heels, she sniffed at the dusty calligraphy of hoof-print and pug-mark; but most of the trails were old. At the edge of her vision she could see the white rumps of Thomson's gazelles as they flitted into the night. From the west came the distant rumble of the Marsh lions;

The Marsh lions roared, reasserting their right to the heartland of the pride . . .

the savagery of their voices told her that they had made a kill, but the wind on the plain carried only the faint smell of spring hares and the stench of hyaena.

Ahead loomed a thin dark band of acacias, marking the course of a lugga. While the lioness sat on the bank, the cubs plunged off along the overgrown bed of the dried-up water-course after a porcupine. The porcupine grunted and rattled its quills. Two of the cubs heeded the warning; the third, more inquisitive, ran in front and stood with one paw raised, as if wondering how best to kill it. But the porcupine, which had confronted innumerable predators in its life, whirled round with surprising swiftness and jumped backwards at its tormentor. The startled cub in turn leapt back as if stung, and the porcupine waddled off up the lugga, leaving one of its quills in the young lion's muzzle.

Shortly before dawn, Young Girl stopped and roared. Her deep groans drummed across the plain, filled the secret unseen luggas, crashed against the ridges, startled impala in the thickets as the echoes sang around them, fading. A pause; then, from out of the darkness, less than a mile away, an answering roar. Young Girl recognized the voice of Shadow, one of her Miti Mbili pride companions. She walked on at a quicker pace while the sky grew lighter in the east, then stopped again, listening intently as a sudden outburst of growls came rolling down from a nearby ridge. Knowing the pride had made a kill, she turned and led her three thin cubs towards them.

Just below the crest of the hill they found Notch crouched possessively across a warthog, watched from a short distance by Shadow and the two-year-olds. Hunger had not improved the temper of the ragged-eared lioness. When the rest of the group had moved in for their share, Notch had turned on them angrily, then had dragged the carcass into a thicket where she tore hungrily at its hindquarters.

Young Girl approached warily, unsure of her reception after her long absence. She walked towards Shadow with tail high, moaning softly as she rubbed her broad forehead against the neck and shoulders of her old comrade. The normally placid Shadow barely acknowledged her greeting and bared her canines in a brief grimace as if to discourage further ceremony, whereupon Young Girl lay down beside her.

Together the two adult lionesses watched as Notch feasted on her kill, pulling the intestines of the dead pig through her teeth to expel the contents before swallowing. From time to time she lifted her bloody muzzle to return their stare with an expression of implacable hostility. Torn between hunger and fear, they hovered, hesitating, knowing better than to approach her now; but the sound of crunching bone and sinew was too much for Young Girl's starving cubs. As they edged forward, Notch rose and lunged at them with a snarl. The cubs scattered in terror. Two disappeared among the rocks; the third went spinning as Notch's flat heavy paw caught him with the full weight of her two hundred and fifty pound frame. At once she was on him, biting so deeply into his throat that he died soon afterwards, choking on his own blood as he crawled away.

Quarrels over kills were commonplace, especially in those lean months. Faced with the threat of starvation, self-interest became paramount; and the lionesses, often so fiercely protective towards their cubs, virtually abandoned their parental responsibilities in a ruthless struggle for survival. Young Girl made no attempt to defend her cubs. Instead, as soon as Notch's back was turned, she rushed forward with Shadow and the two-year-olds to seize what remained of the warthog. Within seconds it was torn apart and Young Girl retreated with a haunch in her jaws.

The sun rose. Light flared along the kindled ridge where the cubs still hid in the rocks. Young Girl and Shadow licked the gore from each other's muzzles. The tension had eased, but a hint of menace still hung in the air. The two-year-olds were nervous and kept apart from the adult lionesses. All eyes were on Notch. Eventually she rose and disappeared into the bushes where the dead cub lay. When she reappeared, carrying the cub by the scruff of the neck in a grisly parody of motherhood, all that remained was the head, skin and forelegs.

Young Girl looked on, seemingly cowed and uncomprehending. From time to time she moaned softly, calling to her cubs, but they were too frightened to return so long as Notch remained. Only when the big lioness had wandered off, followed at a distance by Shadow and the rest of the group, did they emerge from their hiding-place. Even then, long after they had rejoined their mother, she continued to call in the voice that lionesses use when separated from their cubs.

*   *   *

It was now more than a year since the death of Dark Mane, the old Marsh male. The seasons had come full circle. The Marsh pride was secure and the cubs were growing fast. In February, heavy rains swept across the Mara, and the zebra on Paradise Plain enjoyed the flush of sweet grass which sprang up after the storms. The shimmer of green did not last long, but it provided the plains game with much-needed grazing which would help to sustain them until the long rains broke.

In mid-March the male topi began to stake out their breeding territories. The big glossy chestnut antelopes, with their long, sooty faces and their habit of standing motionless on termite mounds for hours on end, were a feature of the Mara. Large numbers lived in the reserve all year round. During the dry season they banded together in large herds as they roamed in search of forage, but, as the time for mating approached, the adult males marked out their stamping grounds with dung piles and much horning and pawing at the earth. So vigorous did these rituals become that Topi Plain was always pitted with shallow muddy depressions, and here the males pranced and strutted in extraordinary postures, the sunlight gleaming on their bruise-coloured haunches as they sought to entice a harem or galloped out to confront a rival.

The graceful impala, too, were re-arranging their labyrinthine social hierarchies as the rival males challenged each other. The impala are animals of the grassland frontiers. Their world is one of glades and groves

The topi pranced and strutted in extraordinary postures . . .

and acacia thickets; wherever there are trees and bushes to provide noontime shade at the edges of the open plains. Now, as dust storms danced along the darkening skylines in anticipation of the coming rains, every morning echoed to the clash of horns as the impala rams competed for the favours of the females.

The rain-birds cried; the rains fell, flooding the rivers, replenishing the dry luggas and bringing the burnt-out plains back to life. After the long rains there was no greener place on earth than the Mara. Once more the reeds sprang tall in Musiara Marsh. On Paradise Plain and all over the great emptiness of Miti Mbili the grass stood green and deep, breaking like waves in long gentle ridges as it rolled away to a horizon so faintly blue and far away that it seemed like the edge of the world. The sound of the grass was in the wind, and its rank vernal smell was everywhere.

The rains washed the dust from the air, so that when the sun returned it shone with redoubled intensity. The light in the East African highlands is not like the soft water-colour skies of Northern Europe. In the high pure air of the Mara, barely a hundred miles from the Equator, it sparkles clear and diamond-bright; and its peculiar clarity, falling across the far-ranging vistas of the open plains, makes for a wild and heady sense of freedom that few visitors to the reserve fail to experience.

In the wake of the rains came the birds. As each fresh storm died away, flocks of Abdim's storks came spiralling out of the sky like dark snowflakes

to stride through the wet grass, stabbing at insects. Where deeper pools had formed, great white egrets hunted for frogs, skewering them on their yellow bills. Time and again, flappet larks rose from the grass and then drifted back to earth, clapping their wings in a rapid tattoo.

Great white egrets hunted for frogs . . .

Sickle-winged hobbies returned to prey on the termites, which they caught in mid-air, snatching them with their taloned feet and transferring them to their sharp beaks with hardly a pause. The hobbies – migratory falcons like small peregrines, with russet leg-feathers and dark, piratical moustaches – had come up with the rains from the Rufiji River, deep in Tanzania, where they wintered. Now they were moving north again, some to breed as far away as the piny heaths of southern England, where they would spend the summer.

Buried in the Marsh the catfish emerged from the muddy burrows in which they had lain protectively entombed through the drought. Their splashings in the reedy pools attracted the attention of fish eagles and saddle-bill storks. All day the Marsh echoed to the gull-like cries of the watching eagles as they flung back their heads and yelped at the sky. One of them glided low over the water and rose with a fish in its talons, but was immediately attacked by another. Together they soared into the sky, twisting and tumbling until the fish fell back into the Marsh and was snatched by a saddle-bill.

Nearby, a huddle of yellow-billed storks were feeding in the shallows.

The fish eagle soared into the sky . . .

One of them caught a large fish and the others rushed towards it with wings raised. In the ensuing free-for-all a grey heron calmly stepped forward and intimidated the yellow-bills long enough to grab the fish for itself.

Where the sharp reeds grew, scarlet dragonflies zipped across the water. Balanced on a dead stump, a pair of pied kingfishers sat and digested their meal. Male and female were indistinguishable except for the extra band of black which the cock bird wore on his chest. Their nest was a tunnel in the muddy banks of the Mara River; there, when the floods had subsided, the female would lay her six white eggs on a midden of fish bones. Now they hunted, hovering over the fen like huge moths, to plunge and emerge with small fishes which they bludgeoned to death against their perch.

Everywhere, reactivated by the rains, insects were on the march. Safari ants poured across the ground in black treacly columns, devouring everything in their path. Dung beetles with glittering blue and indigo carapaces trundled away balls of manure many times their own size and buried their trove in underground chambers to make edible nurseries when their eggs hatched. The beetles in turn became food for baboons and banded mongooses.

The insect world is a microcosm of the greater order of mammals whose savannas, forests and marshes they share. Like the animals, they are both prey and predators. The browsers are the caterpillars, crickets and grasshoppers; the carnivores are spiders, assassin bugs, robber flies and hunting wasps. Theirs is an endless world of mass murder, committed silently and unseen in the green aisles of the grass stems. By swiftness or by stealth, they kill and are themselves eaten in untold millions by the birds and small mammals which flock to the feast.

As always it was the termites which provided the greatest banquet, and not only for ant-bears and scaly pangolins which dug them from their subterranean cities. All over the plains their mounds and mud chimneys erupted, releasing hordes of winged alates into the waiting mouths of chats and drongoes, hornbills, hoopoes and lilac-breasted rollers. So prolific were the termites that besides the more common insect-eaters they attracted bateleur eagles, Montagu's harriers, yellow-billed kites and flocks of European storks.

Released from their dry season refuge in the riverine forests, elephants split up into small family groups and spent the days far out in the grasslands, feeding on the tender young acacia seedlings which they uprooted with coiled trunks, sometimes helped by a nudge of their front feet. They returned to the forest in the early mornings when the grass was still soaked with dew, moving swiftly and easily with a swaying rhythm, the small calves dwarfed by the massive bulk of their mothers.

The oldest among the elephants which came regularly to the Marsh was a veteran cow with a single tusk, the Musiara matriarch. Her life spanned half a century, stretching back to the days when few elephants were found in the Mara. She had been born in the Lambwe Valley, a

hundred miles west of the reserve near Lake Victoria. While she was still a small calf, the Kenya Game Department had sent professional hunters to clear the valley so that it could be opened up for settlement. Hundreds of elephants were shot, but the valley was never settled because of tsetse fly.

Not all the elephants were killed. Many fled across country and eventually found sanctuary in the Mara. Among them was the future matriarch of Musiara Marsh. There she grew up, a fierce recluse, keeping to the dense riparian forests, often pursued by sportsmen and wandering 'Dorobo hunters, and growing increasingly wiser in the ways of avoiding man. Over the years the Mara elephants multiplied until they could be counted in hundreds. The old matriarch herself had produced eight offspring, losing at least two of them during the notorious poaching years of the 1970s when the demand for ivory transformed what had been sporadic slaughter into an international multi-million dollar racket. Elsewhere in Kenya elephants died in their thousands as back-country traders and small-time city crooks vied with corrupt officials and cynical politicians for a share of the bonanza. In northern Kenya their willing accomplices were the *shifta*: Somali bandit gangs armed with AK-47s from the Ogaden war. In Tsavo park the elephants also fell to the poisoned arrows of the Waliangulu and 'Kamba tribesmen. In the Mara the

. . . the small calves dwarfed by the massive bulk of their mothers . . .

The Marsh lions held no fear for her . . .

poachers were mostly Wakuria and Watende who lived on the escarpment outside the reserve.

Now, having survived by luck and by guile, the old elephant was no longer so wild. In spite of the poachers, her years in the reserve had made her more docile and approachable. Slowly, she had become habituated to the presence of tourists and no longer rushed into the forest whenever a vehicle appeared. Instantly recognizable by her single long and slender tusk, she was a familiar sight in the vicinity of the Marsh. Age showed in her sunken cheeks and hollowed temples, accentuating the gaunt planes of her bony head. Her big, idly flapping ears had been torn to ribbons along their lower edges – the result of innumerable encounters with the unforgiving thorns – and her eyes, surprisingly soft and liquid brown, were hooded and enfolded in a mass of wrinkles that gave her an aura of immeasurable wisdom. The Marsh lions held no fear for her but, because she knew the big cats were quite capable of taking a very young calf, she would not tolerate their presence near the herd. If she encountered them in the reeds she would move towards them, ears outspread, stridently trumpeting her displeasure until they slunk away.

During the long rains the constantly dripping trees irritated the old matriarch, and she moved out of the forest to feed in the open. Sometimes she was accompanied by a large herd, comprised of a loose confederacy of

related family units and kinship groups; but more often she was followed only by the dozen most closely related cows and calves of her own family. Altercations between them were few. They seldom used their tusks in anger. Elephants are, in truth, gentle giants who clearly enjoy the close and gregarious nature of family life, and display a remarkable degree of tolerance towards each other. There is no doubting their intelligence. Alone among animals, elephants seem to possess some shadowy concept of death, and the loss of one of their number is invariably deeply disturbing. Just after the long rains, two separate tragedies occurred in close succession among the Mara herds which highlighted this unique quality.

A young calf in the matriarch's own family group fell sick and died one night not far from Rhino Ridge. All next morning the mother stood guard over her dead infant, her big ears flapping backwards and forwards as she tried to keep cool on the shadowless plain. From time to time the tip of her trunk would hover gently over the body of the calf, touching and sniffing the slack mouth, the closed eyes, as if searching for signs of life. When there was no response, she tried to raise it from the ground with her forefoot. About midday she moved off, aimlessly plucking at tufts of grass and sucking water from drying pools; but halfway through the afternoon she returned, making her way unerringly through the acacia bushes to where the calf lay, and stayed until long after darkness had fallen.

. . . gentle giants who clearly enjoy the close and gregarious nature of family life . . .

The tip of her trunk would hover gently over the body of the calf . . .

The three bulls returned, hurrying forward with ears flared and trunks outstretched to gather round their fallen friend . . .

Shortly afterwards, a bull elephant died in the grassy meadows between the Marsh and the River. One of four young bulls which had been frequenting the area for some time, he had wandered away from his mother at the age of fifteen and had eventually joined a small bachelor group, preferring its casual companionship to the solitary existence sometimes chosen by more elderly bulls. This particular group had become inseparable since the rains began, and when he collapsed beside a game-viewing trail at the edge of the Marsh, his three comrades had immediately gathered round in consternation and had tried to raise him to his feet. It had taken him a long time to die. He lay on his side, his great bulk shuddering as he laboured for breath. He made no sound, but his trunk writhed and coiled on the ground in the agony of his sickness. All night he lay, helpless and unable to move while the hyaenas ate into his back and the soles of his feet.

In the morning the three bulls returned from the forest, hurrying forward with ears flared and trunks outstretched to gather round their fallen friend. Once more they tried to help him to his feet, pushing in vain with trunks and forelegs. Long after he died they remained by his side, touching and sniffing as the tips of their trunks played over his body and, in their reluctance to desert him, exhibiting something very close to grief.

8

June-August 1979

The Dogs of Aitong

*They awaited the coming
of the zebra . . .*

RIPE AND drooping under the weight of whiskered panicles, the June grasses had reached their full height when the wild dogs returned. Already the lush swales of Rhodes grass had lost the quick sharp green of the rainy months, the stems growing coarse and hoary as they awaited the coming of the zebra, and the oat-grass had begun to take on the smouldering blood-red glints of summer.

The wild dogs were a vanishing species. Mysterious, elusive, enigmatic, they were the restless corsairs of the wide grass oceans. They possessed no territories, and their sudden appearance would convulse the peaceful plains game into headlong panic and confusion. One day they would arrive where they had not been seen for months; a week later they might be miles away. Only when the bitches whelped would the packs settle down for a few brief sedentary months, hunting in the vicinity of the den until their pups were old enough to accompany them on their marathon journeys.

For years the dogs were despised as vermin, shot and poisoned by farmers, hunters and game departments alike. All over Africa their history had been one of unremitting persecution. Condemned as stock-raiders, blamed for the decimation and dispersal of the wild herds, loathed for the way they disembowel their living prey, they became a fugitive breed, to be eliminated wherever they were found.

Old prejudices die hard; but, as more enlightened attitudes began to prevail, the wild dog appeared for the first time in its true light – not as a bloodthirsty and indiscriminate butcher, but as a highly intelligent social animal whose hunting efficiency actually *improved* the quality of the plains game by removing sick individuals and scattering the herds to prevent inbreeding. Even the gruesome manner in which the dogs dispatched their victims came to be recognized as no less efficient than the frequently clumsy butcherings of lions and cheetahs.

Sadly, the change may have come too late to save the dogs. Despite the protection of parks and reserves, they have continued to decline throughout their range, and there are now probably no more than a few thousand left in the whole of Africa. Even in the Serengeti the packs have dwindled to the level where natural diseases take such a heavy toll that they may never recover – and in the Mara they have reached the very edge of extinction.

Two years before, the Aitong pack had returned to their old dens on the Plain of Stones, with lacklustre coats and bald patches testifying to the poverty of their condition. There were then eleven adults. Within a month of their arrival they had produced fifteen puppies; but two weeks later only five adults still survived and all the pups had died. Exposure may have accounted for some of the losses after unseasonally heavy rains had flooded the dens; but the rest were struck down by some unknown disease. There were no recognizable signs of the most likely killer, canine distemper.

The remnants of the pack, four males and a female, remained in the area for another two months. They were seen one evening pulling down a young wildebeest, but were driven from the kill by the Murram Pit hyaenas. Next morning, however, they killed again, and their dawn hunts became a regular event on the plains where the migratory herds had gathered. In October, one of the males disappeared and was never seen again. Towards the end of the month the survivors were observed early one morning when they flushed a female leopard near Leopard Gorge. They chased her in an almost playful manner as she streaked for cover, but gave up long before she gained the sanctuary of the gorge and

Their sudden appearance would convulse the peaceful plains game into headlong panic and confusion . . .

They followed the departing herds down towards Rhino Ridge . . .

vanished into one of its welcoming fig trees. After that they followed the departing herds down towards Rhino Ridge, were glimpsed briefly on Paradise Plain, and not seen again that year.

By February 1978 they were back at Aitong, where they were observed making an unsuccessful attack on a group of zebra, and afterwards killing a gazelle. They were also seen mating but, although they dug dens, they did not breed. In June, tragedy struck again when another male mysteriously disappeared. Now the need to increase their numbers was desperate, for only three dogs remained: two adult males and a female.

By September the surviving trio had drifted down towards Paradise Plain on the heels of the migration. There they were seen pursuing a young wildebeest, only to be chased from their kill by Scar, who stood on an ant-hill with his mane blowing in the wind, roaring as they trotted away, three lonely figures disappearing into the dusty whiteness of the south.

Now, nine months later, having survived all the rigours of their vagabond existence, they were back, drawn by the same irresistible urge which compelled them to return time after time to the place which held the secret of their own beginnings. Once again their haunting, melancholy voices sang on the wind as they passed through the land of the Marsh lions. With the pack reduced to three dogs, an accident to any one would affect them all. Unlike cheetahs, the dogs were truly dependent on each other and were hard-pressed to survive alone. Their destiny lay in the belly of the pregnant bitch as she trotted north over Topi Plain in the company of the two males. Apart from a few small groups outside the reserve, they were the last of their kind, the remnant of a once prolific presence, and the next few months could decide whether the Mara would ever see them again.

In the Mara the dogs seemed to produce their litters to coincide with the coming of the wildebeest, so that the migration would ensure a bountiful supply of meat for the pups. They did not stay long in the oat-grass country of the Marsh lions. Their breeding grounds were on the drier plains outside the reserve at the foot of Aitong Hill where there was less competition from lions and hyaenas. Here the shorter grass attracted plenty of gazelles, which they hunted until the wildebeest appeared in July.

Wild dogs are the wolves of Africa. Their heads are broad, their muzzles short. When they yawn, their jaws reveal a formidable array of teeth whose jagged edges have evolved to shear through flesh and sinew. There is strength in the muscular neck, stamina in the deep chest, tenacity and endurance in the long slim legs. They are not built like the cheetah to produce blistering sprints. Rather, they are coursers who seldom raise their hunting pace above thirty miles an hour, but can maintain a steady speed for miles, wearing down their prey in a remorseless and single-minded chase to the death. Streaming across the plains in hot pursuit they have a way of running that is beautiful to watch in its effortless economy of movement; lean and easy, but also chilling in its unremitting momentum. The high plains are the home of many killers: lions, leopards, cheetahs, hyaenas, martial eagles; but the dogs are the deadliest of them all.

All three dogs trotted with their distinctive big bat ears cocked to pick up the sounds of the plains. At the head of the trio ran the Bitch. Normally hers was the slenderest silhouette, her belly shrivelled to a lean tightness between the loins; but now she was swollen with pups and her nipples swung heavily as she moved. Even without the signs of pregnancy, she would have been instantly recognizable by her white face markings. Thick white fur sprouted from her cheeks and from inside her ears. Neck and forehead were also white, showing up her dark muzzle and central parting, and there were conspicuous white patches around her ankles. When they hunted, the pattern was nearly always the same: the Bitch in front, followed by the dominant male with the subordinate male in the rear. All three were skilled hunters, but she was the one for the supreme test – the pursuit of the Thomson's gazelles. Compared with these, the wildebeest were easy meat. To run down a gazelle took every ounce of pace, wind and stamina the Bitch possessed. Since her pregnancy, however, she had begun to slow down, and it was the dominant male who now initiated the hunts. There was no mistaking him. Mange or some similar disease had attacked his coat, exposing the sooty skin beneath, so that apart from a thick ruff of fur around the throat and a few odd patches, he was quite hairless. He was known as Black Dog. The subordinate male, too, suffered from bouts of mange, but when he was in good condition he had the most typical coloration of the breed, his body brindled in tortoiseshell hues of black, white and earthy yellow. His shoulders were covered with a pale mantle, and his hindquarters chequered with a jigsaw pattern of white splotches. All three dogs

Apart from a thick ruff of fur around the throat and a few odd patches, he was quite hairless . . .

Early that morning they had killed a gazelle on Topi Plain . . .

displayed the characteristic white tail tips which had given rise to the Maasai name for wild dogs: *Il-oo-ibor kidong'o*, 'those who are white at the tail'.

Early that morning they had killed a gazelle on Topi Plain, then sought the shade of an acacia, where they had licked themselves clean and rested until late afternoon brought a cooling breeze. Now they were on the move again, crowned plovers shrieking about their ears, the light falling gold, the folds and hollows of the distant escarpment filling with soft lavender shadows as the Bitch led them unerringly northwards, where Aitong Hill, its base obscured by the rolling plain, lay hull down on the horizon.

They crossed a shallow lugga and found themselves in a flatter, more open country littered with rocks like charred skulls. They had reached the Plain of Stones, and at once they recognized the familiar candelabra shapes of euphorbias, the patches of Sodom apple bushes and the unmistakable hog's back hump of Aitong Hill itself. Beyond the dusty track that led away to the Maasai townships of Lemek and Narok, the Bitch veered towards a solitary umbrella acacia in the middle of the plain. There she began to cast eagerly backwards and forwards with nose to the ground and wagging tail. Black Dog and the brindled dog stood and watched, intrigued by her strange behaviour. Then she ran towards the slight swelling of an old termite mound and disappeared down a hole.

A century ago, perhaps longer, an aardvark had come shambling over the plain on its regular nocturnal search for termites. Finding the mound, it had ripped a hole in the side and had burrowed deep down through the intricate galleries and fungus gardens to scoop up the seething insects with its long sticky tongue. The following day, when the ant-bear had gone, a warthog enlarged the cavity with his tusks and used it as a sleeping-hole until he fell prey to a lion. But the refuge did not stay vacant for long. Its next residents were a family of hyaenas who deepened it and used it briefly as a den. Then, after they had moved on, it had a whole series of tenants: cobras, monitor lizards, owls, jackals, porcupines, civets and tabby gangs of banded mongooses; until one day a wild dog found it empty and whelped there, when it became what it still remained – a favourite breeding den of the Aitong pack.

Presently, the Bitch emerged from the hole, having satisfied herself that it contained no intruders; then she set to work to prepare the grass-lined chamber in which she would give birth. The two males, meanwhile, had given chase to three hyaenas which had come looming up in the gathering darkness. Although they were more than twice as heavy as the dogs, they lacked the nimbleness and determination of their smaller adversaries, who swiftly sent them packing with quick bites on their rumps. The dogs would not tolerate hyaenas near the den, and in the days that followed, all three would attack any who were foolish enough to come too close.

As her time grew near, the Bitch became increasingly reluctant to accompany the two males when they hunted, preferring to rest at the entrance to the den, though she was still as aggressive as ever towards hyaenas. One evening, towards the end of June, she withdrew into the burrow and was still there next morning when Black Dog and the brindled male left at first light to go hunting. When they returned two hours later, stomachs sagging with gazelle meat, the Bitch emerged from the hole to beg for food, which Black Dog instantly regurgitated; but her stomach was no longer swollen. Sometime during the night she had produced her litter.

Safe in the comforting darkness, the five pups grew fast. After two weeks, their eyes opened, and at four weeks, when they finally emerged, blinking in the sunlight at the mouth of the den, they had already tasted solid food. By now the migration had returned. The plains were black with wildebeest and their calves, and the two adult males were able to provide a regular supply of fresh meat for the Bitch and her family.

The puppies were extraordinary. The females outnumbered the males by three to two, which is unusual for wild dogs. They were still mostly black, not yet having acquired the brindled coats of the adult dogs, and their wrinkled faces gave them a wizened and curiously ancient look. Their ears were too big, and they stank from the day they were born; but – to human eyes – there was also something intensely appealing about their playfulness and the close bonds of affection that bound the pack together.

Whenever the puppies emerged from the den, all three dogs would lick and fuss over them, though sometimes the Bitch would shoulder the other two away if she felt they were acting too boisterously. Every day the youngsters spent longer at the entrance to the den, pouncing, wrestling and chasing each other, or playfully biting the tails and ears of the adults until the patience of the Bitch was exhausted and, with a gentle nip or two, she would roll the squealing offender on to its back.

When it was hot the pups retreated into the den, and the Bitch would follow them to doze in the shade until late afternoon. The adult males, meanwhile, rested under a clump of euphorbia trees some two hundred yards to the east. Only when the sun began to drop towards the escarpment would Black Dog stand up and scratch himself before trotting off to greet the Bitch. His arrival always provoked a frenzy of excitement as the puppies bundled out of the hole to be licked and nuzzled. The Bitch would join in, and then the brindled male, all uttering their characteristic eerie twitterings until the entire family were pirouetting around the mouth of the den in an ecstasy of affection.

On cold mornings, at daybreak, the adults often huddled together for warmth, delaying their hunt until the sun had taken the chill from the air. But on moonlit nights they sometimes hunted in the small hours and would be back at the den before sunrise. To feed the entire pack, the dogs needed to kill at least once a day. Even if they had killed at dawn, the two males would usually set off again in the last hour of daylight to pull down a young wildebeest.

Daybreak and dusk: these were the killing times, the sun blood-red in the smoke of grass fires, the dogs with clean muzzles and thin stomachs which showed they had not eaten, the wildebeest plodding in endless columns along horizons as level as the sea. In these early days around the

. . . the wildebeest plodding in endless columns along horizons as level as the sea . . .

den the Bitch did not leave her puppies, and it was Black Dog who
initiated the hunts, prancing and twittering around his comrade until the
ritual ended and the two dogs trotted silently away across the plain.

With only two dogs hunting, they had, of necessity, become highly
selective in their choice of prey. The zebra families were too well guarded
by the flying hooves and savage mouths of the stallions. Even adult
wildebeest were too difficult to seize and disembowel, but calves were
another matter. A short charge was usually enough to separate a victim
from the herd. One dog would grab the calf by the hind leg or the
tail; the other would go for the throat, and it would all be over in
a matter of minutes. There was no swift *coup de grâce*, but the calves died
quickly and in deep shock.

The dogs fed hurriedly, gulping down gobbets of meat as if they, too,
found the butchery distasteful. Black Dog was always the first to finish,
leaving the brindled male to eat alone. By the time his companion had left
the carcass, Black Dog had already regurgitated food for the Bitch and
her pups. Now, as the brindled dog arrived at the den, chest and muzzle
dark with gore, the pups rushed to meet him, and the Bitch begged with
them, whining and licking until he, too, produced a meal for them.

Sometimes a grey Land-Rover from the Mara River camp would draw
up quietly nearby as they fed outside the den. The dogs paid it no
attention, so well accustomed were they to the presence of tourists, but the
inquisitive pups would follow the vehicle until the adults called them
back. Some of the watchers viewed the act of regurgitation with distaste,
failing to recognize it for what it was: the simplest and most efficient way
of carrying meat from kill to den without the risk of losing it to other
predators.

Humans on foot caused the dogs to react very differently. The first
glimpse of a Maasai herdsman would cause the Bitch to bark and
hurriedly shepherd her pups into the den. As they grew older, the pups
themselves began to recognize the signs of danger. They learned to fear
the roars of lions. They hid from the flitting shapes of hyaenas. If the
shadow of a martial eagle swept over the grass towards them, they no
longer needed their mother's prompting to go to ground. But this last
lesson had been hard won. Martial eagles are hunters of small mammals,
magnificent birds whose black heads and ermine chests give them a look
of barbaric splendour, like medieval knights. One was often to be
seen in the vicinity of the den, where he perched at dawn in the flat-
topped acacias and solitary wild olive trees, glaring across the grasslands
with gold-rimmed eyes. One day, as the pups were playing in the sunshine
outside the den, one of the big raptors came rocking and swaying over the
ground towards them. They cowered in the grass as the hiss and thrum of
the eagle's pinions beat about their ears, but the Bitch, too, had seen him
coming. Even as he lowered his talons, she had leapt to meet him.
Thwarted, he had checked his rush and lofted skyward with a yelp of
disappointment as her teeth snapped beneath him. It had been a narrow
escape, but it showed the pups something they would never forget.

The Bitch was still heavy with milk but she no longer bothered to lie down to give suck. Although the pups would not be fully weaned until they were three months old, they were already quite large enough to reach their mother's teats by standing on their hind legs. When they were eight weeks old, she began to leave them on their own and rejoin the males in the hunt. Almost immediately she assumed her old position of initiator and leader. The wildebeest had begun to drift away from the Plain of Stones, and the two males seemed content to acknowledge her superiority in pursuit of the gazelles and impala which now became her prime targets.

The weeks of relative inactivity around the den had not blunted her hunting edge. Once she had picked out her prey, it seldom escaped. Her first kill came after she had run at full speed for half a mile, when she caught her prey by the hind leg and expertly flipped it on its side before darting in to rip at the belly. The killing was mercilessly efficient. Even if the male gazelles tried to double back to avoid her, they would only find themselves outflanked by Black Dog and the brindled male.

The shadow of a martial eagle swept over the grass towards them . . .

The most remarkable aspect of their hunting, apart from its undeniable bloodiness, was their complete lack of aggression towards each other. When wild dogs feed, there is none of the vicious squabbling that characterizes lion and hyaena kills. The adults willingly stand aside and allow their youngsters to feed first, for the hunting ethic of the dogs demands fair shares for all, and a pack will hunt again rather than allow any of its members to go hungry.

In August the dogs abandoned their den and moved to a new site, half a mile to the east, on a low hill crowned with scattered euphorbias. Here they remained only a few days (having discovered belatedly that the Maasai often used the track at the foot of the hill) and took possession of another den still further east on a level sweep of stony plain. Why they left the original site was a mystery, but their behaviour seemed to indicate a growing restlessness, as if they were fretting to be off again on their wanderings.

Now, even though game was still abundant nearby, they hunted far beyond the Plain of Stones, often ranging as far south as the blond grasslands of Miti Mbili to harass the marching wildebeest. Awakening at dawn, they would stretch and shiver in the cold grey light while the pups slept on in the den's rancid warmth, and the Bitch would begin to frisk around Black Dog until, aroused by her frenzied twitterings, he would begin to prance beside her, his ears raised like enormous butterflies. Tails wagging, they would trot over to the brindled male, who would fawn at their feet and present his neck in a gesture of appeasement before he, too, was swept up in the weird arabesques which always preceded a hunt. Then, still grinning submissively, he would follow them out into the dew-soaked brightness of the awakening savanna.

Coqui francolin and red-necked spurfowl scurried from their path or took flight in an explosion of wings. Where a passing shower had raised green shoots, herds of gazelles were feeding, nuzzling at the ground with heads bowed as if in worship. As the dogs gambolled past, every head went up. Some of the territorial bucks began to pronk, bouncing away in stiff-legged alarm; but the pack ignored them and loped on over the horizon, where wildebeest were emerging in an endless column from the morning mist.

The dogs ran on towards them without increasing their pace. All their senses were sharp, but they hunted by sight, and they shared with the cheetah the uncanny knack of spotting the slowest and weakest animals in the herd. Imperceptibly they changed their gait as they drew closer. Gone was the casual, romping stride. Now there was no mistaking the hunting pose: muzzles forward, ears laid back. Tension crackled across the plain. Zebra thudded away in a cloud of dust. The wildebeest halted, grunting, fearful, watching the dogs with clownish faces. Then, one by one, they galloped past the oncoming pack in a desperate game of Russian roulette. The dogs waited, singled out a mother and calf, and cut them out of the fleeing herd. The calf bleated. Bravely the mother charged Black Dog and sent him sprawling, but he was up in a flash, snapping so fiercely that she

quickly gave up and cantered away.

The Bitch, meanwhile, had grabbed the young one by the tail. With the mother now driven off, Black Dog spurted forward and sank his teeth into the calf's throat, while the brindled male ran in under the hind legs and opened it where it stood. Moments later, all three dogs were straddling the carcass, the hamstrings taut in their thin hind legs as they thrust and wrenched. Soon nothing was left except a bag of wet skin and, an hour later, that too had gone, a trophy to be fought over by darting jackals. All drama forgotten, wildebeest and zebra grazed peacefully again as if nothing had happened. Doves purred drowsily in the morning sun, and five pups slept with full bellies at the mouth of their den.

Black Dog sank his teeth into the calf's throat, while the brindled male ran in under the hind legs . . .

*   *   *

Early one morning, not long after the pack had moved the pups to the euphorbia country, the adults surprised a leopard and her cub near the new den. The cub was a yearling male and, although two-thirds as big as his mother, was still dependent on her. The two cats had been stalking guinea-fowl, but the wary polka-dotted birds had caught a brief glimpse of the cub's whiskered muzzle as he slid through the grass, and had flown off to a nearby lugga where their rusty voices still cackled in indignant alarm. When the leopards saw the dogs loping over the plain, they flattened uneasily in the grass. To be caught in the open in broad daylight was a situation to be avoided. The dogs halted, alert and stiff-legged, ears cocked and muzzles thrust forward towards the spot where the cats had disappeared. They had seen the glimmer of a stealthy body and sensed the presence of a large carnivore.

Minutes passed; the leopards remained hidden, their spotted coats invisible among the stippled shadows of wind-shaken grass-heads; but eventually their insatiable curiosity got the better of them. Cautiously the female looked up to see where the dogs had gone, her muzzle, flat head and laid-back ears so cunningly levelled that she barely broke the surface of the waving grass. The dogs had not moved. Silently she sank back out of sight; but all her guile had been wasted. The watching dogs had seen the smouldering eyes, the dark blunt nostrils aimed like gun-barrels, and began to move forward.

The leopard fled, the cub at her shoulder, matching his mother stride for stride as they streaked for the lugga with the dogs in full cry. At first, the two cats easily outpaced their pursuers, but they were quickly winded. Their bodies, designed for sprinting, could not suck in enough oxygen to sustain tired limbs over long distances. Fifty yards from safety their fluid strides began to falter. The cub had now fallen behind its mother but, as the dogs began to draw closer, both cats reached the sanctuary of the lugga and disappeared.

For a while, the three dogs sniffed among the bushes; then, satisfied with the outcome of the chase, they trotted back to the den. It was a fortunate ending. If the leopards had turned and fought, they could have inflicted terrible damage with their slashing claws. With only three adults and the pups still too young to survive by themselves, it showed once again how slender was the thread by which the future of the Aitong pack was suspended.

9

September 1979

Summer Slaughter

The pride had enjoyed a glut of kills . . .

BULGING with meat, the Marsh lions sprawled contentedly in the long grass. Later, as the day grew hotter, they would drag themselves off to seek the shade of the fig tree forest at the southern end of the Marsh, but in these chill early hours, when the plains glistened with dew, they welcomed the sun on their tawny flanks.

Ever since the migration had returned in July, spilling across the Mara River in a wave of spectacular crossings, the pride had enjoyed a glut of kills. Even now, the carcass of their latest victim, a wildebeest bull dispatched at daybreak, lay farther out on the plain, seething with vultures. The tension all the prides had felt when prey was scarce had lifted with the arrival of the Serengeti herds. The three Marsh sisters and their five large cubs had all fed together on the dead bull with only minor squabbles, though eventually the majestic figure of Scar had come padding through the soaking grass to seize his share. They had eaten until their bellies sagged, then had walked down to the reeds to drink before flopping down in the sun. Now they lay in an untidy heap with Scar resting a short distance away in aloof and shaggy splendour.

The cubs were growing fast. The helpless kittens which the lionesses had suckled in the fig tree forest were now the size of leopards and had been fully weaned since the long rains. Soon they would lose their milk teeth and acquire the formidable canines of adult lions, though their coats were still marked with the shadowy ginger spots and rosettes of juveniles.

They still relied on the adult lionesses to provide them with kills, but they were learning to recognize the taut postures and alert facial expressions which indicated a hunt, and now rarely spoiled a stalk by miaowing or romping ahead of their mothers. Instead, they responded at once to the stalking crouch, sinking down and dropping back as the lions fanned out towards their quarry, and rushing forward only when the hunt was over.

Their own hunting instincts were revealed in many ways. When food was scarce they had shown less inclination for play but, since the wildebeest had returned, they romped boisterously at dawn and dusk, pouncing on each other with exaggerated leaps and wrestling in mock combat; or they chewed sticks and chased francolins through the grass. More recently they had discovered a new game. Every morning banded mongooses emerged from their burrows in old termite mounds. They foraged in packs up to thirty strong, twittering and churring to each other as they hunted for elephant and buffalo droppings, which they would break open in search of dung beetles. The sight of a family of these inquisitive small animals snaking through the grass was irresistible to the cubs. Once their curiosity had been aroused, they had to investigate; quickly discovering that mongooses could do them no serious harm, they pursued them at every opportunity.

At the slightest hint of danger the mongooses would stand on their hind legs, balancing on their rudders with front paws dangling to whistle a warning before racing for the nearest burrow. The cubs quickly became adept at mongoose hunting and, once they had learned that their quarry could not run so fast in long grass, there was no escape. Mongooses were bowled over in mid-stride or flattened with a paw. Some died quickly, bitten through the body. Others, less fortunate, became playthings to be tossed in the air, tugged and fought over until the cubs tired of their sport and left them to the kites and vultures. They never ate the mongooses themselves, apparently finding the flesh distasteful.

Two hours after sunrise, the Marsh sisters left Scar in the long grass and wandered off towards the trees with the cubs close behind. When they reached the edge of the forest they stopped, hearing the crack of branches and dull belly rumbles of feeding elephants. The lionesses did not care to have their midday siesta disturbed by the Musiara matriarch and her family, and sauntered towards the dirt road that led to the Governor's Camp airstrip.

When they were not occupying the Marsh itself, the pride often drifted south to visit two other favoured haunts. Both had been created artificially during the construction of the road, and the lions had been quick to discover their usefulness as places of ambush. One was the abandoned murram pit, which had become a permanent water hole for plains game and which now had a resident hippo and a thriving community of dabchicks, herons and Egyptian geese; the other lay midway between the camp and the airstrip, where the road crossed the southern end of the Marsh on a low embankment. Concrete culverts had been laid under the road to carry away the floodwaters that built up after the rains, but the water was slow to drain away in the low-lying black cotton clays, and luxuriant reed-beds sprang up every year following the rainy season. For the lions it was a perfect refuge. The marshy pools attracted thirsty prey to be ambushed from the reeds; and on hot days the big cats often crawled into the cool culverts to sleep.

Tourists, returning to camp for a late breakfast after their early morning game-viewing drive, passed the Marsh pride padding along the road, but did not pause to see them drop down the bank towards the

They romped boisterously at dawn and dusk . . .

BELOW LEFT: *Every morning banded mongooses emerged from their burrows in old termite mounds . . .*

The cubs quickly became adept at mongoose hunting . . .

culverts. Had they waited they would have seen the first lioness warily sniffing at the entrance in case the pipe was already occupied. If there had been a lion inside, it would almost certainly have been another Marsh lion, probably one of the Talek twins or their offspring, for the Marsh pride was now at the height of its power. This was its heartland, and no solitary nomad or infiltrator from a neighbouring pride would have dared to lie up in the culverts and risk being trapped by the sudden arrival of Mkubwa, Brando or Scar. Nevertheless, the lioness was cautious. She sometimes surprised hyaenas here, and had to chase them out of the culverts before she and the cubs could enter. This time, however, the pipe was empty; but the air inside was musky with the odour of a marsh mongoose.

The mongoose lived and hunted in the adjoining pools and reed-beds, where it ate frogs, lizards and small birds. It was larger than the banded mongooses of the dry grasslands, sharp-nosed and glossy with a longer tail and rich chocolate fur which caught the light with an oily sheen. It was a night hunter whose insatiable appetite often drove it to forage by daylight, and it had come snuffling through the culvert only moments before the Marsh pride's arrival. One by one the lions crept through the tunnel, emerging at the other end, where the scent was strongest, to wait while the leading lioness moved stealthily towards the reeds. With discovery now inevitable, the mongoose made a sudden dash for freedom, but there was no real chance. The lioness was on it in an instant, and strutted back to drop it at the feet of her cubs.

The mongoose was not badly hurt. It stood its ground, puffed out its fur and spat with rage whenever one of the young lions reached out a paw; and it cut short the cubs' favourite cat-and-mouse game by darting forward and sinking its sharp little teeth into an over-inquisitive muzzle. The cub jerked back, shaking and pawing at the small brown head; but the mongoose hung on and did not release its grip until another cub ran forward and bit it through the ribs.

Afterwards the pride stayed in the vicinity of the culverts, preying on the zebra when they came to drink. But within a week they were back in the Marsh, where sunrise found them resting beside two wildebeest they had killed in the big reed-bed below the spring. One carcass was partially eaten, but the other lay untouched. The Marsh sisters seemed disinterested in their kills, as if they were already sated. Yet even so, when one of them spotted a slight movement at the entrance to a pig-hole, she went at once to investigate. Her haunches wriggled as she began to dig, sending up spurts of dry earth as she clawed her way deeper into the burrow. She kept on furiously scrabbling until her head and shoulders were completely hidden inside the hole; then suddenly stopped and backed away with a *woof* of surprise. As the dust cleared, a large mongoose shot out past her and dashed into the Marsh.

Unlike its gregarious banded cousins, the white-tailed mongoose was a solitary animal. By day it slept curled up in abandoned warthog burrows, emerging at night to catch snakes, guinea fowl, cane rats and ground

squirrels. It belonged to the biggest of all the six species of mongoose which live in the Mara, but its only defence was an evil-smelling secretion squirted from glands under its tail. The lioness, crashing through the reeds in close pursuit, grimaced as she caught the full force of it, allowing the mongoose to escape unharmed.

In the heat of the chase the Marsh sister had failed to notice that another lioness had been drawn by the commotion from the reeds. Now, as she turned to rejoin her companions, she saw the newcomer and recognised her as one of the Talek twins, whose cubs – the Marsh twins – were developing into a handsome pair of young lionesses and were fully integrated into the pride. Already they had played their part in ambushing and killing their first wildebeest calves. Four months earlier the Talek lioness had mated with Scar; and now a second litter lay concealed among the reeds. The new cubs were barely two weeks old, and the Talek mother had been about to move them to a new hiding place when she heard the mongoose dash past her in the reeds. She growled softly when she saw the Marsh sister, twisting her ears to show the black markings and warn her pride companion not to come any nearer.

Afterwards the Talek mother gently closed her mouth around one of the cubs and carried it into the fig tree forest. Twice more she returned to the Marsh, and each time she emerged with a tiny cub dangling from her jaws. When three cubs had been safely deposited in the forest, she went

Each time she emerged with a tiny cub dangling from her jaws . . .

again to the birthplace in the Marsh and spent a long time searching in the depths of the reeds, occasionally uttering the hollow moaning grunt of a lioness summoning her young; but, when she finally reappeared, her mouth was empty.

*   *   *

The cubs did not stay long in the fig tree nursery. Within ten days of moving them from the reeds, the Talek mother took them to a new hideout in the riverine forest to the west of the Marsh. Game was abundant along the river and she could hunt on her own until her family were old enough to be introduced to the pride.

The grassy bays and marshy meadows bordering the forest were drying rapidly. Out in the Marsh the reed-beds shrank before the trampling hordes of wildebeest. Where buffalo had sucked and shuddered to their wallows in the black cotton mire, the ground was hard as cobblestones. Thousands of hungry mouths had nibbled it bare. Weeks of drought had baked it to a pitted crust. Now the very earth itself was being stripped away as the dust devils twirled in the heat-shimmer. The land did not dry evenly. Pans and depressions still held a flush of green where moisture remained, and the park-like glades with their wide-spreading fig trees teemed with animals. Here lived the creatures of the forest edges; bushbuck and waterbuck, buffalo and impala. Topi stared from old

The reed-beds shrank before the trampling hordes of wildebeest . . .

Five giraffes saw her go . . .

termite mounds, whose shallow domes rose from the seed-spangled grasses like the burial mounds of some long-forgotten necropolis, and troops of baboons shuffled among the trees like arthritic old men. The big dog-faced males barked in alarm as they spotted the sway-backed outline of the cheetah Nefertari, drifting across a wide clearing where gazelles grazed in small groups. Within a month of losing her cubs to the Marsh lions, she had mated with Badfoot, the limping male, and was pregnant again.

All this the Talek lioness observed as she skulked in the riverine thickets. Early each morning, when the sun lit the tops of the tall forest trees and the turacos began to call, it had become her habit to lie in wait at the edge of the woodlands. Shadows hid her. She lay relaxed, her hind legs spread out on one side; but her eyes were alert and her head was up, with her weight on her elbows. In front of her was a shallow depression which during the rains had held a large pool. Though most of the muddy bottom had long since become a dustbowl, water still glistened where elephants and buffalo had deepened it with their wallowings, and warthogs loved to recline in its soothing black ooze.

As yet the wallows were deserted, but the warthogs had already emerged from their burrows to dig for tubers at the edge of the clearing. Silently the lioness stepped from the shadows. Five giraffes saw her go, a lean tawny shape stealing through the grass towards the pigs. Volleys of doves hurtled overhead, their wings hissing in the windless dawn, but the

Violence hangs about them, in the vicious curve of the heavy horns . . .

lioness did not look up. Hunger drove her forward. She craved the hot salt taste of blood, and all her attention was on the rooting pigs. Whenever they lifted their snouts, she froze in mid-stride, gliding forward again as they returned to their foraging. Soon she was close enough to charge but, as she bunched for the final rush, she was distracted by a cow buffalo moving slowly across her line of vision, heading for the Marsh.

Buffalo feed mostly by night. In the dry season they tend to stay close to the Mara River, waiting until late morning before they emerge from the forest to lie up in their marshy wallows. So it was unusual to see a buffalo wandering so early towards the reed-beds, and rarer still to find a cow by herself, though small groups of bulls are common. Suspicious, short-tempered and cunning, they are formidable adversaries. Violence hangs about them; in the vicious curve of the heavy horns, in the truculent lift of the muzzle, and in their stance – foursquare and threatening – dark nightmarish shapes with mud-caked flanks, sifting the air with fierce flared nostrils. The bulls are particularly aggressive, and the lions respect their strength. But a solitary cow is a different proposition; and a sickly one is an invitation which no hungry lioness can ignore. As the labouring animal drew closer, the Talek mother forgot the warthogs and veered away to intercept.

The cow's progress was painfully slow. Drained by the parasites that swarmed inside her, she plodded with bowed head, her tired hooves kicking puffs of dust from the pool's dried bed. Yet, as soon as she caught the scent of the lion, she turned to face the danger. She even gathered herself for a half-hearted charge, but the lioness side-stepped with ease and caught her from behind. Iron claws dug into dung-caked rump and both animals crashed to the ground. At once the lioness released her grip and straddled her victim like a wrestler, straining for a suffocating bite across the muzzle. But every time she edged forward, the cow struggled and tried to rise. Her bellows echoed across the Marsh, causing eight pairs of round tawny ears to twitch and lift. Moments later, the Marsh sisters and their cubs came trotting from the reeds.

The numbers contined to grow until, an hour later, there were thirteen lions around the kill. They included the young Marsh twins, the Talek mother's barren sister and the regal presence of Scar. Only the two other pride males, Brando and Mkubwa, and the Talek mother's three new cubs were absent. It was one of those rare occasions when almost the entire pride was together. They stayed near the carcass for the rest of the morning, crawling to sleep with bloated bellies in the shade of the reeds; but their peace did not last long. Just before midday they were disturbed by the main buffalo herd on its daily trek to the Marsh. Vultures were squabbling over the kill, and the buffalo approached warily. Suddenly one of the herd bulls scented the lions and wheeled round with muzzle held high. Others joined him, snorting angrily, then together they advanced in a tightly-packed mass. The young lions bounded away in fright as the closed rank of iron-grey horns swept down towards them; but Scar and his lionesses had seen it all before. They had seen lions tossed like sacks and pulped under the hooves. In their own early lives they had seen litter-mates die with stomachs opened by single sweeps of those terrible horns. But they were confident of their own superior speed and agility. When they retreated they gave ground grudgingly, moving no faster than they had to until the bulls lost interest.

The lionesses melted away into the grassy wastes of Topi Plain, but Scar did not follow them. Now that he was fully fed he had no need of their company, and went his own way. The buffalo had not chased him far, but his heavy meal and the pulsating heat had made him thirsty. He found a pool in the reeds and crouched beside it, lapping rhythmically at his own reflection.

Scar was now a magnificent lion. The gaunt and nervous nomad of two years ago had grown into a sleek and self-assured pride male. From his battle-scarred nose to his black tail-tuft he was nearly ten feet long. The rangy body had filled out and his mane was so thick that it almost hid his ears, a glossy rug of luxuriant tobacco-coloured hair, shot through with auburn glints. He was now in his prime, broad-muzzled, deep-chested, immensely powerful, and his tenure as a resident pride male was reflected in the magisterial, almost insolent swagger of his stride as he rose from the pool and padded into the shade. Of all the three Marsh males, Scar was the most possessive, the most territorial in his habits. While Brando and

Scar was now a magnificent lion . . .

Mkubwa often went off together for days at a time, sometimes trying to re-associate with the Miti Mbili lionesses, Scar preferred to stay close to his own pride. He liked the cover of the reed-beds, the shelter of the forest edges. He liked to drink from the same spring-fed pools, to sleep through the incinerating noons under the same shady figs. In his regular patrollings, renewing his scrapes and scent-marking his boundaries, he had come to know every inch of his territory. He knew every game trail, every glade and lugga. Sometimes he ranged far out into the open grasslands of Topi Plain, or south beyond the airstrip lugga as far as Rhino Ridge; but his true home was the Marsh. Its familiar sights and sounds and smells reassured him, and he was content to remain there so long as his lionesses continued to use the reed-beds for their kills.

When they were hungry, all three males would converge on a kill. But afterwards, when they rested, it was rare to find them under the same tree. Brando and Mkubwa had always tended to consort together, but Scar seemed quite content to be alone. Perhaps he felt less need to seek alliances. Often, when darkness fell, all three would go their separate ways. One might pick up the scent of a nomadic lioness. Another might be attracted by the sound of hyaenas on a kill, or might investigate the roarings of other lions. Their relationship with each other was of the most casual kind, and at times their role as pride masters appeared strangely inconstant and obscure. They were not leaders. Instead, they were content to follow the lionesses and feed off their kills. But their collective presence gave the pride stability and protection, and the fact that there were now three sets of healthy cubs was a measure of their strength.

Now, bathed in shadow, Scar slept. He lay on his side with his mane rumpled around his shoulders, occasionally stretching to ease the not unpleasant lassitude that seemed to weigh down his limbs. Sometimes his eyes would open and he would look about him; but all he saw was a silence of grass, and the crouching hills beyond. Out on the plains the air quivered and scintillated along the stony ridges. Nothing moved. Everywhere animals rested, seeking refuge from the bludgeoning heat. Each pool of shade held its patient herd of impala. Warthogs withdrew into their burrows. Buffalo lay in their wallows. Elephants moved deeper into the forest, fanning themselves with flapping ears. Marooned in the hot silence of the open grasslands, giraffes stood motionless beside solitary acacias, and fiscal shrikes called with maddening insistence from the somnolent noontide luggas. Their scolding voices did not disturb the resting lion. Apart from brief and exhausting bouts of mating, most of his days were spent in slumber. Patrolling his territory or following his lionesses were mainly nocturnal activities. He lived from one meal to the next; from the zebra and wildebeest pulled down by his pride, to the sudden violence of his own kills when he was lucky enough to surprise an antelope or bowl over a warthog in the long grass. In between, he slept. He was not lazy. Rather, he conserved his energy, moving only when compelled to do so by hunger, or by sex, or by the need to assert his territorial authority.

Two hours before sunset, a new breeze came sighing over the grass. Scar awoke and lifted his head. As the sun slid down, animals emerged from the shadows and began to feed again. Zebra and topi moved slowly along the horizon. A stormcloud bloomed above the escarpment, bearing rain from Lake Victoria. Above it the sky became an immense apple-green cupola, clear and empty except for the specks of orbiting vultures. The cooling air revived him. Leisurely he stood up and walked out into the plain. The grass enveloped him, closing behind him as he moved through the gently bending stems. He felt the brush of its plumed seed-heads along his flanks, smelled its dry odour crushed beneath his heavy pads as he sniffed at the sharper taints where Thomson's gazelles had scent-marked individual spikes of grass with special glands on their cheeks. But when he paused to test the wind, lifting his muzzle above the tops of the grasses, he heard the familiar grunting of wildebeest and padded silently towards the fig tree forest. The trees lay across the lower end of the Marsh in a ragged crescent. Normally they provided excellent

Panic spread across the
Marsh . . .

cover which enabled the lions to approach the reeds unobserved; but, as
he slipped through the leafy shadows he was seen by a troop of Syke's
monkeys. Their coughs of alarm alerted the wildebeest so that, when Scar
stepped out into the open, they were already scattering before him in a
rush of tossing heads. Panic spread across the Marsh. Birds flew up,
crying wildly. Zebra and topi burst from the reeds and splashed to the
safety of the open plains. The lion ignored them all. Not until he found a
wildebeest trapped in the mud did his body tense with anticipation.

As the dry season advanced, the Marsh had developed treacherous
boggy patches. Churned to a thick morass by constant trampling, the
shrinking pools had become soft quagmires, and it was all too easy for
stampeding wildebeest to blunder into them. Scar was not hungry. He
had fed well on the buffalo his lionesses had killed that morning; but the
prospect of such an easy meal was too tempting to pass by, and the
wildebeest's frantic struggles only served to sharpen his predatory
instincts. He waded forward, submerged to his belly, until he was close
enough to reach out a muddy paw and grab the wildebeest by its head. He
closed his great mouth over its muzzle, twisting the animal on to its back
and holding fast until the twitchings ceased. But the Marsh was reluctant
to yield up its treasure. The heavy body had sunk too deep, and Scar's legs
slipped and skidded in the quaking mire as he struggled to haul it from the

mud. The muscles bulged under his mane, but the wildebeest would not budge. In the end he gave up and left it for the scavengers. There would be other wildebeest; and, as long as he remained with his pride, there would always be lionesses to kill for him. Marsh ooze dripped from his jowls as he strode into the plain. In the dying light his mud-caked body was black and full of menace. For a moment he paused, the rug of his mane billowing against the sky. His roar crashed through the falling dusk and boomed across the Marsh. Bushbuck at the forest edges heard him a mile away, and moved off in alarm. Then the grass closed around him and he was gone.

Scar's legs slipped and skidded in the quaking mire as he struggled to haul it from the mud . . .

*   *   *

Out on the plain where the two trees stood, the lioness Shadow was resting with three new cubs. She, too, could hear Scar roaring as she lay on a rocky rise overlooking Miti Mbili lugga. Her pride still lacked resident males, but that did not mean she was unable to mate, for nomads and amorous males from neighbouring prides would always seek out the Miti Mbili lionesses when they were receptive. So it had been with Shadow. Courted by the two Marsh males, Brando and Mkubwa, she had mated with them both.

Shadow had seen little of the Miti Mbili pride since the cubs were born.

Ten weeks earlier the group had moved away to hunt along the northern limits of their range and Shadow, tied by the arrival of three tiny cubs, had been unable to participate in their kills. However, two days previously they had followed Notch down the Miti Mbili lugga, drifting south with the wildebeest as the herds grew restless for the Serengeti. Shadow, too, had become increasingly restive, and was eager to rejoin the pride. During the night she had heard lions roaring. The voices were those of her companions, Notch and Young Girl; and, as soon as it was light, she led the cubs westward across the plain to look for them. For the cubs, padding behind their mother on short wobbly legs, it was the first real journey into open grassland. Until now they had been kept hidden among rocks, sheltered by thorny thickets or concealed in the shadows of scrub-choked luggas. Wide-eyed they watched the migrating flocks of Caspian plovers flickering low over the land. Startled, they jumped back in bewilderment when a shorthorn grasshopper whirred into the air from beneath their paws to glide away on quivering mauve wings. The bright treeless plain was another world, new and exciting, but also filled with danger. There was no cover from predators. Their only protection was the reassuring presence of their mother as they followed her black tail-tuft across the close-bitten sward.

Wildebeest watched them file past. They knew the lioness was not hunting and did not run, but every head turned to face her. The cubs blinked in alarm at the strange, grunting creatures. This was their first encounter with wildebeest, and the sight of the horned herds sent them scampering after their mother. At first the cubs had followed eagerly, pausing to cuff each other in play and then galloping to catch up. But, as the sun grew steadily hotter, they began to flag, yowling plaintively as they trailed farther and farther behind until Shadow stopped to wait for them. On these occasions she would lick and nuzzle them, gently cajoling and encouraging, sometimes even picking one up, though they were now almost too big to be carried and objected strongly whenever she closed her mouth around them.

In this manner the strange little procession continued across the plain towards the distant line of Miti Mbili lugga. Wildebeest were drawn in their wake, trailing after the lioness as if they could not bear to let her from their sight. Their movements were seen by hyaenas of the Miti Mbili clan, resting half a mile away in an isolated patch of croton scrub. Sensing a meal, they stood up in ones and twos and stared intently into the swimming heat-haze. When the four pulsating blobs materialized into a lioness and three small cubs, their fur bristled, their tails rose stiffly erect and they came rocking over the grass to investigate.

Moments later, Shadow found herself confronted by fifteen hyaenas. She stopped and sat on her haunches with the cubs cowering between her forelegs. Her ears flattened against her head and she pulled back her lips, exposing her canines. A cavernous growl rattled from her throat, but the hyaenas were not impressed. A single lioness was no match for the combined strength of the Miti Mbili clan, and they knew it. They fanned out in a half-circle, grunting and laughing with open mouths as they closed in for the kill. No longer was their mother's bulk a sufficient reassurance to the cubs. Squirming from beneath her belly, they ran and tried to hide behind her. The hyaenas rushed forward. Whenever they tried to outflank her, the lioness lunged and threatened, keeping them at bay with sudden low rushes, but she could not hope to defend her cubs for much longer.

It was pure chance that saved them. A driver had spotted them through binoculars and led a procession of three tourist vehicles rushing towards them over the plain. Clients always wanted to be shown lions, and this morning until now the *simba* had been unusually elusive. The hyaenas paused, and the momentum of their attack was lost. Their aggression subsided and the mob scattered, leaving Shadow to comfort her cubs. She remained there with them for perhaps fifteen minutes, staring anxiously after the retreating hyaenas, then called to her family with a low moan and resumed her journey.

Later that day the driver found her again. She had finally tracked down the pride on Miti Mbili lugga and was lying contentedly between

They were now almost too big to be carried . . .

Her flanks rising and falling to the easy rhythm of her breathing as her cubs played and suckled . . .

Notch and Young Girl with the cubs at her side. In the grass a hundred yards away, vultures were hunched over the remains of a wildebeest killed by the lions during the morning. Young Girl's two yearling cubs were also there, together with the six sub-adults which Notch and Shadow had raised between them. All the lions were resting amicably and it was clear that Shadow's new cubs had been fully accepted into the pride. Shadow herself seemed none the worse for her encounter with the hyaenas. Her eyes were closed to amber slits. Her body shone lithe and tawny with sunlight, her flanks rising and falling to the easy rhythm of her breathing as her cubs played and suckled.

Lions die many deaths. Some perish of starvation or disease. Many die violently, gored by buffalo or killed in crippling fights with their own kind. A dead lion does not remain for long. The beneficiaries of the bush – hyaenas, marabous and vultures – see to that. Shadow was never seen again. If she died, her body was never found, and her disappearance remains a mystery.

10

October-November 1979

The Spotted Assassins

Couched in the fork of a monumental fig tree, a leopard slept, waiting for darkness. She lay as limp as a rug, her body draped along a branch, all four legs dangling. From time to time her eyes opened, pale discs of glazed green, the pupils contracting to black pinpoints as the breeze laid back the leaves and splashed her face in sunlight. Even when her eyes were closed, the small round ears twitched and turned, sifting the sounds of the hot afternoon; the keening kites, the bubbling cries of coucals, the tremulous twittering of mousebirds in the leafy canopy. Hers was the semi-sleep of cats, in which the senses rarely fall below a level of subconscious awareness.

The tree she had chosen was the oldest of all the wild figs in Leopard Gorge: massive and wide-spreading, grotesquely gnarled. A century ago, at about the time the first Europeans set foot in the Maasai country, the sweet fruit of its parent tree had been eaten by a baboon. The hard sticky seeds had passed unharmed through the animal's gut and had lodged in the crotch of a wild olive. There, one seed had germinated. Roots sprang from the seedling fig, feeling their way down the trunk of the host tree until they reached the ground. The fig grew fast, and the olive supported its growth. Greedily the parasite wrapped itself around until, in time, the olive tree was choked and the fig took its place on the rim of the gorge.

Leopards are solitary – they live wherever there is cover . . .

Since then it had sheltered many generations of leopards, and its silver-smooth bark was deeply scarred from the rake of their claws.

This leopard was one of two cubs born in the gorge nearly two years ago. She would soon be fully independent, though she continued to share her mother's home range. Her father was the huge wall-eyed male who roamed over a much wider territory and had killed two of Young Girl's new-born cubs on Miti Mbili lugga the previous year. His one blind eye had done little to impair his hunting prowess. He was a master of concealment, a denizen of the dark and secret thickets. Silently he would slip under arches of thorns, cleaving the grass as softly as smoke. The wary impala never heard the tread of his midnight paws until his powerful hundred and fifty pound frame struck them with gasping force.

Sometimes the young leopard saw him on his nightly prowlings, a sinister shadow drifting under the trees; or heard his ripsaw cough hack through the darkness. Once she had almost walked into him as he sunned himself at the edge of the gorge. She looked up, feeling his unblinking stare, and saw him watching from behind a boulder; the broad muzzle, the one good eye agleam, the other milky and opaque.

In the tree she felt secure. Here she could rest, high beyond the reach of the Gorge pride lions. Leopards and lions feel a mutual antipathy. They are competitors, fellow carnivores sharing the same hunting grounds, but with little else in common. Lions are social cats, animals of the wide savannas, and their coats are the colour of Africa itself. Leopards are solitary, and shun the plains. In the Mara they live wherever there is cover; in the rocks of kopjes, in lairs of thorns, in the riverine forests and shady luggas, and on the overgrown slopes of the Siria escarpment. They are the nightwalkers, moonlight hunters whose spotted coats render them almost invisible in dappled shadow.

Only an hour earlier, a Land-Rover had come lurching down the gorge. Its occupants had been scrutinizing every tree, but still they had failed to see her no more than thirty feet above their heads, so perfectly did her coat blend with the play of light and shade, hiding the thick creamy fur of chin and throat with its necklace of spots, the rich saffron coat and its blotchy rosettes. She was not a big leopard. Even when fully grown, she would probably weigh no more than eighty pounds – about one third the size of a lioness. Her shape, too, differed from that of a lion. Her tail was longer in proportion, and her neck much thicker, making the spotted head look small and snake-like.

A few years earlier, her chances of surviving long enough to raise cubs of her own would have been slim. Leopard-skin coats were still fashionable in most Western capitals, and the cats were still being ruthlessly trapped and killed. After ivory and rhino horn, leopard skins were the most prized trophies to come out of Africa, and, as the demand grew, poaching became highly organized. In Kenya the poachers often used baits laced with cattle dip. The poison was cheap and easy to obtain in Nairobi. Its only disadvantage was that poisoned leopards often died slowly, far from the bait. Sometimes they were never found; or found too

An eagle owl caught the sudden movement . . .

late, when the fur was ruined. Nevertheless, shooting and poisoning were so effective that, by the mid-1970s, leopards were hard to see, even in the Mara, which had been a major stronghold. Firm action by the Kenyan Government in 1977, coupled with a growing world-wide revulsion that human vanity could endanger an entire species, probably saved the Mara leopards. The young female in her fig tree was living proof of their revival.

From her airy perch she could survey the entire length of the gorge. Beyond the figs and cactus-like euphorbia trees which sprouted from its rim, the ground rolled away in a stony, tsetse-infested waste of whistling thorns until it joined the open plains. Even from so great a distance, the leopard could see that the grasslands were devoid of game. Within the last two weeks the herds had left for the Serengeti. Once more the vast concourse of wildebeest and zebra had gone south in a cloud of dust, leaving the Marsh lions to hunt pig and buffalo, or to scavenge kills from the Murram Pit hyaenas. But the seasonal movements of the plains game scarcely affected the leopard. Unlike the lions, whose lives swung between glut and famine, she could depend on an inexhaustible supply of smaller prey. She preferred impala and Thomson's gazelle; but she would also eat hares, dik-dik, hyrax, genets, rock pythons, guinea fowl – even the bright metallic-blue starlings whose chattering throngs sometimes shared her larder trees.

Once the sun had plunged behind the escarpment, night came swiftly in the eastern sky. The leopard stretched; her claws unsheathed. With forelegs flattened and hindquarters raised, she scratched the bark until it bled. Then, turning easily in her own length, she leapt to a lower branch and balanced, black against the sky, her long tail curved in an elegant question mark. From its perch in a tree on the opposite wall of the gorge, an eagle owl caught the sudden movement. The owl came regularly to the gorge to kill hyrax which lived in holes amongst the rocks. A faint sound behind it caused the bird's head to swivel sharply. When it turned back again, the leopard had gone. She had come down the tree in a series of zig-zag bounds, her paws barely touching the bark as she flowed soundlessly into the shadows.

The moon rose; a hunter's moon, ripe and round and yellow. In the clear air of the high plains its light fell with a hard and searching brilliance. In the distance the bleached October grasslands shone silver like the sea, crossed by the ghostly shapes of feeding animals. Night sounds; frogs and cicadas, hyaena, zebra and, from the south, from Musiara Marsh, the groan of a lion. All this the young leopard perceived as she prowled on the dark side of the gorge. A soft breeze drew intriguing scents across her nostrils as civets, ant-bears and other night creatures emerged to snuffle along the game trails. Tourists never saw them, but in the mornings their nocturnal wanderings would be revealed: a maze of prints, punctuated by fan-like flurries where nightjars had pressed their wings, and the pug-marks of the leopard which had failed to catch them.

Just before dawn, however, she managed to surprise a guinea-fowl, and carried it to a tree where she plucked the feathers fastidiously from its breast before eating. Soon afterwards she saw another leopard entering the gorge; she stiffened, but relaxed when she recognized her mother and dropped from the tree to greet her. The two cats did not yet avoid each other, as they would when the daughter reached maturity. Instead they walked side by side, pressing flanks in their affection, the daughter extending her tail like a friendly arm over her mother's back. The mother in turn responded by licking the fur of her daughter's neck. But afterwards they went their separate ways; the mother to a cave in the walls of the gorge, the daughter to her tree.

She recognized her mother and dropped from the tree to greet her . . .

She leapt to follow them . . .

Hyrax were basking in the sun's first rays by the time she reached the fig. From cliffs and crevices, dozens of pairs of beady eyes watched her unhurried progress through the gorge. Their cries of alarm echoed among the rocks as she sniffed round the base of the tree for signs of other leopards, but the only scent was her own. She was still hungry, but now she would rest. She sank back on her haunches and launched herself upward. For a second she clasped the vertical trunk like a wrestler and hung there, spread-eagled, six feet above the ground, every muscle straining as her claws dug into the bark; then, in a series of clinging bounds, she sprang lightly into the upper canopy and stretched out along a branch. She stayed there all morning. Later, when the day became overcast, she came down from the tree and lay in the rocks at the edge of the gorge. Shortly before dusk the cloud cover dispersed and the sun flared briefly over the escarpment, flooding the gorge with an angry light. At the same time a breeze arose, bringing with it the scent of game, and the tiptoe click of hooves.

Picking her way over the stony ground was a female impala with a small fawn. The leopard froze until they disappeared briefly behind a thicket; then she leapt from her rock to follow them. Sometimes her thick neck and short stumpy legs could make her look almost squat; but now she was all fluid grace. She did not run; she poured through the grass, belly to

ground, tail held low, her head flattened like a snake's. Some sixth sense seemed to warn the impala. She stopped, wide-eyed and alert. Normally she lived in a herd, where many eyes made it harder for a predator to approach undetected. But a few days earlier she had left to give birth to her fawn; now she was leading it to a resting place above the gorge. Suddenly uneasy, she snorted and skittered off a few yards, followed by the fawn, and wheeled to stare where she sensed the presence of eyes. But the leopard had moved again, using the cover of the thickets, and was now less than twenty yards away. Yet even from so short a distance, in the yellow grass, among the thorny shadows, she was invisible.

The fawn lay down in a tussock; and once more the leopard wriggled forward. The ground between them had been burned by the Maasai, and was still bare and blackened, with no cover except for a few scattered rocks. But so guileful was the leopard that she was already midway across the empty space before they saw her. Too late the impala mother snorted and danced away, bounding in ten-foot leaps through the thickets. Too late, as her panic subsided, she ran back, snorting frantically, as if to draw attention to herself and save her offspring. For an instant the fawn struggled to rise from where it had been pinned; but the leopard only tightened her grip on its throat. When its head hung limp, she straddled the body with her front legs and began to drag it away into the rocks. The

The impala mother snorted and danced away, bounding in ten-foot leaps through the thickets . . .

She, too, walked away and vanished in the gathering twilight . . .

impala watched until the leopard disappeared, while the sky turned red. Then she, too, walked away and vanished in the gathering twilight.

The leopard reached the fig tree only a yard or two ahead of three hyaenas, which had come loping in pursuit out of the dusk. But it was now that her neck muscles revealed their true strength and purpose. She sprang into the tree with the fawn in her jaws as if it were no heavier than a bird. An hour later, having swallowed the viscera and chewed at the thighs, she left the carcass and came down to lie on a rock. At once the three hyaenas reappeared as if from nowhere and began to circle. But the leopard ignored them, even when one of them walked right up and thrust its muzzle a few inches from her face to sniff at the blood on her whiskers. It was a measure of her growing maturity that she neither fled nor turned on her tormentors. She was confident of her ability to defend herself if she had to; but she was also a solitary huntress and could not afford to be injured in senseless fighting. Unlike the lions, she belonged to no pride whose kills would feed her if she were hurt. A disabled leopard is a starving leopard, unable to hunt until its wounds have healed.

Long afterwards, when the hyaenas had slunk away, she returned to her kill. There, disturbed perhaps by the encounter, she transferred the corpse to another tree and fed from it again before leaving the gorge to explore another part of her range. She did not return for several weeks, but for many days afterwards the dried remains of the dead impala dangled their gallows warning.

*   *   *

Elsewhere, in the rough riverine meadows to the west of the Marsh, other spotted cats were busy. Servals prowled through the lank grasses, pouncing on francolin and harlequin quail, and a trio of sub-adult cheetahs took up residence, to the delight of visitors at the tented camps. These youngsters, two females and a male, were the offspring of a beautiful ring-tailed cheetah whose range overlapped Nefertari's. When Nefertari forsook the glades to give birth to her new litter on Topi Plain, the three sub-adults moved in. Born in March the previous year, they had been independent of their mother for barely a month, and would remain together as a sibling group until after the short rains, when the young females would come into season for the first time.

The last traces of their juvenile ruffs had almost disappeared. They were more lissome and slightly shorter than full-grown cheetahs, but they had acquired all their mother's hunting skills, and lacked only the finesse that experience would teach them. Fanning out through the tall grass, they had already learned how to stalk their quarry, taking advantage of the numerous ant-hills to avoid flushing their victims until the last moment.

Sometimes, however, they were rebuffed by the most unlikely ad-

Servals prowled through the lank grasses . . .

. . . a beautiful cheetah whose range overlapped Nefertari's . . .

versaries. On one occasion a family of warthogs chased them away, and promptly lay down in the shade which they had been occupying. Another time, they chased a gazelle from the ridge near the spring to the edge of the Marsh. The gazelle was sick, the cheetahs not really hungry, having already made one kill earlier in the day, and their pursuit appeared to be more playful than serious. When they caught it they merely toyed with it, pawing and tripping it as they would a fawn, but never going for its throat. Suddenly, as the gazelle bleated its distress, the reeds parted and a waterbuck appeared. It was the Marsh monarch, the big shaggy male with the record-sized horns. The glade was his territory. Here he lived, feeding on coarse grasses spurned by other grazing animals, or resting in the Marsh, invisible except for the majestic spread of his horns. The Marsh lions seldom troubled him, though he would move away when they were near. He was a heavy animal, the size of a Scottish stag, and the secretions which waterproofed his donkey-brown coat gave off a strong musty smell. Although the pride would occasionally attack waterbuck, they would never do so when there was other game to be had.

Hearing the gazelle, the Marsh monarch had left his drinking pool to investigate. Now as he ran from the reeds, the cheetahs backed away. At first they retreated only a few yards, but when he moved threateningly forward again, they slunk away and took refuge amongst the anthills some thirty yards distant. All this time the gazelle never moved. It had ceased its bleatings and lay in the grass with its chin to the ground like a new-born fawn. When at last the big buck disappeared into the reeds, the cheetahs cautiously returned to their prey. The gazelle waited till the last minute before scrambling to its feet, only to be pursued and felled by one of the females. Her sister joined her, and together they capered around their victim until, tiring of their play, one of them bit across its throat.

The gazelle's long death rattle reached the ears of the waterbuck. Once more he emerged from the reeds, joined this time by a herd of twenty females. The noise of the dying gazelle seemed to arouse and inflame them. They were both curious and agitated, jostling forward with the Marsh monarch at their head until they were no more than fifteen yards away. At first the cheetahs crouched behind their kill, growling with bared canines. Then the young male's patience snapped. He charged forward, scattering the waterbuck, who wheeled and crashed away into the reeds. By the time he returned, padding through the grass with his tail curled over, his sisters had already begun to feed.

*   *   *

Heat poured over the land. In the aftermath of the migration, the plains were stubble. Tyre tracks carved in the black cotton mud of May had baked to rutted iron, and the few pools that remained in the luggas were green and stagnant. The Maasai came into the reserve, seeking water for their bony cattle. With robes flapping, they strode like Old Testament prophets beside their thirsty herds, the white dust boiling in their wake.

Dawn still broke fresh and clear, as it so often does in the high country;

but by mid-morning a restless wind arose. There was no moisture in it; no hint of rain. It was akin to the desert winds that blow in the lands of the Samburu, the Turkana and the Rendille, hundreds of miles to the north, where the rocks are a fiery Martian red and the mountains are hulks of long-dead volcanoes, receding towards Ethiopia. It scorched the stony ridges until they were too hot to touch, scattered the white pierid butterflies like discarded petals as they flew in the grass at the forest edges; and sometimes, when vultures and marabous descended in slow spirals from the sky, it carried the hot thick stench of death where lions had killed in the night.

Far out in the hazy brilliance of the plains, a solitary wild olive stood on the horizon. Generations of browsing giraffes had clipped its high canopy into a neat black parasol whose flimsy shade now encompassed eleven recumbent lions. Farther off, unmindful of the sun's brassy glare, a group of zebra stood, resting their heavy heads on each other's backs as they stared with quizzical eyes into the heat-shimmering distance. They had seen the pride arrive, and knew the lions were still there. From time to time they would see the irritable swish of a black-tipped tail. And sometimes, when the wind died and the heat seemed more intense than ever, they could hear the cats' harsh breathing as they panted in the shade. But the zebra sensed that the lions were not hungry, and did not move. Later, however, when the pride began to stir and stretch under the tree, the stallion squealed at his mares and they cantered away together in the deepening light of late afternoon.

Notch sat up and looked around her. By now the Miti Mbili pride had changed so much that it was scarcely recognizable as the one which had once held the rich hunting grounds of Musiara Marsh. Of her original companions, only Young Girl now remained. Old Girl had been dead for more than a year. And Shadow, too, had disappeared, leaving her three small cubs to perish. With no other suckling lioness to nurse them, they had grown steadily weaker until in the end they had been abandoned to the hyaenas. Yet the group had held together, and the remaining youngsters were thriving. Young Girl's two surviving cubs were now fourteen months old, and the seven cubs born to Notch and Shadow in 1977 had become true sub-adults, rangy and powerful.

The males in particular were beginning to look quite handsome. They were now at an age when, in a true pride, they would have been regarded as potential rivals by the older pride males. But, because there were no resident breeding males among the Miti Mbili lions, the youngsters had not been driven out. This unusual state of affairs added to the tensions within the group and, on one occasion, had produced the extraordinary spectacle of Notch attempting to seduce two of her nephews. The young males were thoroughly bewildered as the normally temperamental lioness suddenly became playful and kittenish, rolling in front of them and sinuously rubbing herself against their flanks. Her persistence eventually stimulated the first awakenings of the mating urge in them, but they were still too young to meet the demands of fatherhood. From time to time, in

. . . the Marsh monarch, the big shaggy male with the record-sized horns . . .

They cantered away together in the deepening light of late afternoon . . .

response to Notch's growing desperation, one male would try to mount her, hindquarters thrusting ineffectively without actually coupling. But when at last his aunt moved away, he slumped in the grass as if relieved to be rid of her tiresome importuning.

Eventually, Notch's desire had been exhausted by waiting. Hunger was again paramount, and her keen eyes scanned the plains. Suddenly her expression changed. Her companions, always quick to react to her mercurial moods, noted the new alertness in her face. One by one, their heads lifted from the grass to follow her unblinking stare.

Notch had spotted the orphan rhino whose mother had been shot by poachers. The young rhino preferred the secrecy of the luggas, the leafy tunnels among the hilltop patches of croton bush, and the arid stony places where the wait-a-bit thorn protected her with its fish-hook barricades. But there were certain herbs and other plants, greatly favoured by rhinos, which only grew on the plains, and it was these which had tempted her into the open, to pluck at the tough but succulent stems with her thick prehensile lips. Being alone did not bother her. Like all black rhinos she was essentially solitary by nature. Unlike the gazelles and zebra which moved freely across the plains, she stood heavy and earthbound in her grassy firmament, a dull-witted dreadnought from an older world. Everything about her was prehistoric: the saurian snout with its spike of horn, the wrinkled hide, the dim and lizard-lidded eyes. Even her strange, three-toed footprints were like the fossil spoor of a dinosaur.

At first she was unaware of the approaching lions. She was standing downwind, and her broad nostrils failed to pick up their scent. But her sense of hearing was uncannily acute. As the nearest young male stalked towards her, she snorted suspiciously and whirled round to face him. He crouched motionless in the grass where, to the myopic rhino, he was no more than a dark shadow. But then the wind changed. The rhino caught the lion smell and charged. The young male sprang aside, but she turned with surprising agility and stood, weaving her head from side to side as she hunted for a scent of him. Soon he was joined by the other sub-adult males, and together they had great sport, stalking up behind her as she tried to trot away, and bounding off whenever she turned to confront them.

The rhino was lucky. Had the lions been hungry they might have tried to kill her. Instead, they were just inquisitive youngsters who showed their inexperience by turning the encounter into a game.

Notch, meanwhile, had taken no part in the incident. Once she had realized that the calf was old enough to defend itself, she had wandered off with the rest of the pride down to Miti Mbili lugga, where zebra were barking in the twilight. Experience had taught her that rhinos were dangerous, cantankerous creatures with tempers even more uncertain than her own.

*　　*　　*

Everything about her was prehistoric . . .

By late October the Marsh was a mere ghost of its former self. The reed-beds were smashed and trampled. Even the water that bubbled perpetually from the spring in the rocks had dwindled to a trickle. One by one the pools had dried until only a few muddy shallows remained where elephant and buffalo had tusked and trodden in their daily excursions from the forest. To these last sanctuaries, fouled by the dung of drinking animals, came egrets, herons and yellow-billed storks, gathering in increasing numbers to fish the receding waters. Every day the sun went cartwheeling overhead through skies of burning blue to quench itself in the smoke of the grass fires along the escarpment walls. And every day the pools grew smaller. The Marsh was dying for want of water.

The land ached for rain. Beyond the Marsh, becalmed in an infinity of yellow grass, giraffes stared across the plain, as if waiting for the brief erratic November showers that would clear the air and raise new grass from the old year's dust. To a casual eye it was a serene and shining landscape, as peaceful as an English park. There was no malice in it, no hint of suffering or hostility. Orioles called with clear voices from the dappled shade of forest figs. Hippos chuckled in the river, and bou-bou shrikes chimed their monotonous xylophonic responses from the heat-drugged thickets. The sounds of summer lulled the senses; but the world of the plains animals was a constant paradox. Nothing was ever what it seemed. Tranquillity was an illusion behind which stalked old familiar spectres: hunger, thirst, disease. The golden vistas, outwardly so innocent and benign, were full of sudden, violent images. The pristine plains were a charnel house of skulls and bones, half-eaten zebra, bloated vultures. Hidden in the tall grass, slovenly hyaenas raised gory muzzles from a shipwreck of ribs, and hungry lions tore at their kills with paws encased in gloves of blood.

When night fell, lightning flickered along the horizon and thunder rumbled in the distant hills. The short rains were coming. The animals felt it, and the waiting made them restless. The Talek mother sensed it as she nursed her cubs in the riverine forest, but her growing unease was not due merely to the sultry weather. Ever since she had isolated herself from the pride to give birth eight weeks earlier, she had become increasingly aware of the presence of other lions. The newcomers were nomadic strangers, outcasts from the Talek prides to the east, and footloose males who had come north with the wildebeest. They feared the intimidating figure of Scar, whose nightly challenge reverberated across the Marsh; but they did not disperse.

In the morning, anxious for the safety of her cubs, the lioness led her family back to the Marsh. Startled waterbuck ran off as she approached, high-stepping through the grass, then stopped to turn and stare once they realised she was not hunting. Cautiously she crossed the clearing, her big flat paws raising puffs of dust which made the cubs sneeze and grimace in her wake. At the edge of the forest she found a pool which still held water, and paused to drink while the cubs wriggled to lick the drops from her whiskers.

Skirting the fig tree forest she came at last to the two ancient figs that grew at the edge of the plains. There she sniffed at the base of the biggest tree, exploring the bark for news, and found Scar's pungent presence still

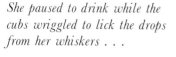

She paused to drink while the cubs wriggled to lick the drops from her whiskers . . .

fresh from the previous night. Then, satisfied that no alien lions were lurking in the immediate vicinity, she slumped in the shade, and the three cubs, exhausted by their long march, snuggled against her and fell asleep at her nipples. Neither she nor the cubs had seen the watchful leopard concealed in the canopy of the fig tree forest.

He was a handsome animal, not as old or as heavy as the wall-eyed male of Leopard Gorge. His ears were not yet scabbed and battle-scarred; his nose was still a youthful pink and his whiskers stood out like porcupine quills. In colour, too, he was strikingly different from the old wall-eyed tom, whose spots and rosettes were a dusky olive brown, like the colours in an old tapestry. His own markings were distilled from a deeper darkness and blossomed as soft as soot on his lustrous coat.

Now he lay in his tree and waited, dozing intermittently; one forepaw supporting his head, the other hanging down. There was no hurry, and his patience was infinite. Occasionally his eyes opened and his chilling stare would rifle through the curtain of leaves. During the afternoon he awoke to see Scar resting with his lioness. At first the cubs were awed by the formidable presence of their father. Later, when they had overcome their fear, one of them even tried to play with him. The big lion did not growl, but merely raised a crooked lip; the sight of the huge canine tooth was enough to send the youngster scurrying back to its siblings. All this

. . . the watchful leopard concealed in the canopy of the fig tree forest . . .

The cubs were awed by the formidable presence of their father . . .

the leopard observed as the sun passed down and the shadows of his forest spread like a dark stain towards the sprawling lions.

There was no moon, but the stars were bright enough for the leopard to see Scar melt away through the grass when other members of his pride began to call from the north of the Marsh. The Talek mother also heard them and gave an answering roar. She could wait no longer. She needed to be a pride lioness again, to hunt with her barren sister and surround herself with the muscular presence of her pride companions when she rested. The cubs were still too young to accompany her into the night, where hyaenas were already whooping. When they tried to follow, she rebuked them with a grimace of displeasure and moaned insistently until they crawled beneath a fallen tree to await her return.

The leopard watched her go, then slipped noiselessly from his branch, hugging the trunk with his belly as he descended head first to the ground, cushioning his landing with the thick pads of his feet. By now the stars had vanished, swallowed up by looming cloud; but the leopard knew exactly where the cubs were hidden. Stealthily he sought them, sniffing under the dead tree until he had found the first one and silenced it with one swift bite. The other two whimpered at the whiskered mask and cowered deeper out of reach. Suddenly, the leopard paused. Perhaps his keen ears heard the lioness returning. For a brief second longer, the cubs remained transfixed by his baleful gaze. Then he was gone. Soon afterwards it began to rain, fat heavy spots that washed the blood from the dead cub's fur and obliterated all traces of the distinctive pug-marks leading into the fig tree forest.

11

December 1979-January 1980

The Nomads

The wildebeest came in their tens of thousands until the land was black with their bodies . . .

SCAR lay alone on Topi Plain, a stone lion in a grey dawn, his grave visage fixed on the distant contours of the Marsh. All around, other lions were roaring. To the east he could hear the Miti Mbili lions. From the Marsh came the familiar groans of his own lionesses; and from the south, the deeper grunting chorus of the Paradise pride males. But recently, too, he had become aware of other, unknown lions whose sullen voices challenged his supremacy.

The year was coming to a close; the hierarchies of the prides were changing. Old alliances were dissolving and new partnerships were being forged. For two years, the Marsh pride had proved invincible. Their territory was without equal in the Mara, yet they had held it with a fierce tenacity. Now new generations had emerged, tough young bachelors impatient to possess prides of their own. Every night they ranged back and forth between the established territories, as Scar had done in his own youth, probing for signs of weakness, waiting for an opportunity to seize control.

As usual, it was the annual migration with its vast influx of plains game which had lured the strangers to the Marsh, The wildebeest came in their tens of thousands until the land was black with their bodies, and the bellies of the Marsh lions sagged under the weight of meat. From dawn until dusk the plains shook to the drumming of hooves, to the grunting voices and sudden uproars that signalled the successful ambush. But the Marsh, whose bounty drew the wildebeest, also attracted the unwelcome attentions of nomadic lions. Most were outcasts from neighbouring prides, three-year-olds caught in the unfamiliar thrall of sexual maturity and eager to mate for the first time. Some had strayed from the Talek prides and other big groups which roamed in the east of the reserve, while others had come from the acacia woodlands of the Serengeti, having crossed the rivers when the waters were low to follow the northward surge of the herbivores.

Amongst the newcomers there were always a few shiftless females, usually from large prides where no new lionesses could be absorbed, but the majority were males. Some of these would never know the security of pride life. They would wander until they died, killing when they could, scavenging from hyaenas, cheetahs and hunting dogs, and forever trying to avoid the hostile prides on whose preserves they trespassed. The nomads came singly, or in twos and threes as brothers or litter-mates. Sometimes these sibling groups banded together for extra protection. One such alliance, of seven young males, remained for several weeks in the no-man's land of Rhino Ridge – the dividing line between the territories of the Marsh lions and the Paradise pride – before retreating across the Mara River to try their luck in the Triangle.

So long as the Marsh pride was protected by the combined strength of Scar, Brando and Mkubwa, the outsiders stood little chance. Together,

The Marsh pride was protected by the combined strength of Scar, Brando and Mkubwa . . .

they were more than a match for their younger rivals. All three were now in their prime: confident, cunning and many times blooded in battle. Of late, however, Mkubwa and Brando had shown renewed interest in the unattached Miti Mbili lionesses, and had begun to spend more time away from their own territory, leaving Scar as sole defender of the Marsh. There was no compulsion on the two philanderers to remain with their own lionesses. They were not bound by ties of loyalty. Like all lions they were opportunists; and the social ethic of their pride amounted to little more than the satisfaction of mutual needs. By establishing territories and defending them vigorously against other lions, the pride males create a sense of order. While it lasts, the lionesses are able to raise their cubs without fear of constant harassment. In return, they become both mates and huntresses for their guardians; for food and sex are the twin hungers which motivate the males and hold them constant in their own domains. Sometimes, however, they encounter nomadic females, or meet unattended lionesses from adjoining prides, and seek to establish new liaisons. In their absence, the ever-watchful bachelors move in, and become increasingly difficult to dislodge.

So the war of nerves continued. Scar remained vigilant. Alarmed by the roaring nomads, the Marsh lionesses closed ranks. The newcomers, for their part, bided their time, nervously rising at the sound of Scar's thunderous challenge – knowing they were trespassers, yet reluctant to retreat. They were still young lions, unsure of their strength. The voices of the Marsh males carried an unmistakable warning, but the young bachelors were held by their own physical needs and desires; not just the promise of easy hunting, but by some deeper urge to assert themselves as pride lions.

Among the newcomers was an alliance of four young males, probably brothers or litter-mates who had stayed together and were now seldom seen out of each other's company. They were typical nomads, nearly four years old, with big muscular bodies, sparse manes and relatively unscarred muzzles. They were such a distinctive group that they quickly became known as the *Wageni*: the strangers.

Hunger kept the Wageni constantly on the move. One day they would appear on Miti Mbili Lugga; next day they would be seen prowling along Leopard Lugga, or furtively nosing around the northern end of the Marsh. Unlike Scar, whose lionesses provided him with a steady flow of kills, the nomads had no pride females of their own to rob. They were forced to scavenge and battle with hyaenas over carrion, to dig warthogs from their burrows and waste their energies in hopelessly misjudged rushes after zebra and topi. They were clumsy hunters, and their lack of stealth and cunning drove them to attack buffalo. The wild black cattle were less fleet of foot than other plains game, but also infinitely more dangerous and difficult to kill.

By mid-November the Wageni were firmly entrenched in the Marsh lions' domain. Miraculously, they had managed to avoid a direct confrontation with Scar, but a trial of strength sooner or later was

A small group of buffalo were feeding and wallowing . . .

inevitable. One evening, at sundown, the four young lions stood staring intently towards the Marsh, where a small group of buffalo were feeding and wallowing. In the swiftly falling dusk, they stole through the grass and hid themselves on the lower reaches of Leopard Lugga. The lions waited until the fading light hid the buffalo from their view, and there was only the occasional splash of a heavy body in the reeds to tell them the herd was still there. This was the best time for the lions to hunt. Often in late afternoon they would awaken and watch, with mounting hunger, as herds of plains game grazed in full view. But they had learned that pursuit was difficult until darkness came. Only then would they prepare their ambush, placing each broad pad with a slow and deliberate precision, as they eased forward silently towards their prey.

Eventually the buffalo emerged from the Marsh and began to move towards the lugga. And still the lions waited as the herd drew nearer, sniffing the air with suspicious muzzles. It is a popular misconception that lions consciously choose to hunt against the wind. Often the Wageni would hunt downwind, only to rise in bafflement when their intended victims picked up their scent and escaped. But this time the breeze was in their favour. Almost before the buffalo were aware of the danger, the cats had burst from their hiding place and the leading lion had hooked his claws into the rump of a trailing cow. Almost simultaneously, the second and third lions hit the buffalo from the side and sent her crashing, while the fourth grasped her by the neck and clamped his teeth across her nose and mouth. By the time she had lost her struggle for breath and he could release his hold, his three companions were already tearing ravenously at rump and loins.

Inevitably, the sounds and smells of carnage brought hyaenas loping in from the plains. Soon the night was full of hunched, misshapen forms as half the Murram Pit clan whooped and circled around the feeding lions. For hours the air was made hideous by the din. Lions and hyaenas charged and counter-charged, lunging and growling for possession of the carcass. Next morning the buffalo had been completely dismantled, with all its legs and ribs wrenched apart. At one point the Wageni had retreated, leaving their spoils to the hyaenas. But two of them had returned, driving off the few remaining scavengers, to feed on what was left. They ate hurriedly, raising their heads between each mouthful to stare nervously around them as somewhere, not far away, the Marsh lionesses began to roar their triumphant cadenzas at the dawn.

Scar, too, was ill at ease. Increasingly he was torn between the need to remain near his lionesses and his urge to drive out the nomads. Night after night he patrolled his territory, reasserting his ownership by visiting all his regular scrapes and scent-markings, until hunger compelled him to look for his pride. But the Marsh lionesses were no longer so easy to find. They, too, could sense the rising tension as pride males and intruding nomads manoeuvred for dominance, and they seldom remained in the same spot for long. They behaved like fugitives in their own territory, and Scar went hungry when he could not find them at their kills.

The task of guarding the entire range was becoming too much for Scar on his own. Some of the younger challengers had retreated without a fight, backing down the moment they glimpsed his massive frame trotting towards them. But the four Wageni were still there. Recently, too, Scar had begun to be aware of two more arrivals. These latest newcomers were not inexperienced sub-adults, but a pair of mature lions whose heavy pug-marks could only belong to large males well into their prime. Their presence added to his growing discomfort. He was tired but he did not sleep. Oblivious to the heat, he would lie out in the open grasslands, not bothering to seek the shade, twitching irritably at the crawling flies and gazing stonily into the distance with eyes that seemed to close less often. Unconsciously he missed the support of Mkubwa and Brando – still pursuing their amorous adventures on Miti Mbili Plain. For the first time in two years, his confidence had begun to waver. The nomads were all around him; their spoor was everywhere. But he was a pride male. This was his territory, where he belonged, and he would not relinquish it easily.

As the day came alive, flights of doves hurtled through the sharp highland air. The first rays of the sun lit their vinous breasts as they raced low over the land. Scar heard the rush of their wings. He heard the joyful songs of white-browed robin chats and morning warblers, the booming of ground hornbills; but his ears were cocked towards the Marsh, which lay still bleary with mist in the middle distance. At last he hauled himself to his feet and began to walk towards it. In a little while he came to a solitary bush and began to draw himself sinuously around it, rubbing his head against the foliage and straining upwards to sniff at the higher branches. Then, turning abruptly, he raised his tail stiffly and drenched the bush with his own pungence, obliterating the scent of other lions

which had sprayed there in his absence. He walked on through the quickly drying grass. Zebra fled, scudding away with shrill barks of alarm; snorting topi cantered from his path, panicking the gazelles, whose white rumps turned this way and that like shoals of fish. Fear spread outward like ripples in a pool. All across the plain, the herds parted to let him through.

Only when he roared did they leap up in surprise and whirl to face him . . .

A light breeze blew across him, lifting the thick bush of his mane as he walked. Sometimes he paused to sniff at the grass, or to listen with eyes half-closed in concentration. Against the light his shaggy crown of tufted copper seemed to ripple with flame. As he drew closer to the Marsh he began to roar. Away to the east beyond Miti Mbili lugga, Brando and Mkubwa heard their pride companion and answered him in unison; the distant thunder of their response was like his own voice returning. Somewhere the nomads would be listening; but, if they heard, they did not take up the challenge.

Silently now, Scar padded along the edge of the reeds towards the spring. Again he stopped, his muzzle wrinkling as he read the wind. The smell of the dead buffalo came to his nostrils; the foetid odour of hyaenas; and something else. Intermingled with the rich stew of scents – and growing stronger at each stealthy step – came the unmistakable taint of lions. Engrossed in their kill, the two Wageni did not see the big pride male appear from behind the reeds a hundred yards away. Only when he roared did they leap up in surprise and whirl to face him. Still roaring, he continued to walk towards them; then came at them determinedly with heavy mane swinging. One of the nomads fled at once, bounding away into the lugga; but his companion was less fortunate. He was the smallest of the four young males and his more powerful companions constantly intimidated him on kills and robbed him of the choicest morsels. As a result of one such skirmish he now walked with a pronounced limp.

Slowed by his injury, the nomad immediately went down under the fury of the onslaught. In vain he turned submissively on his back, writhing and snarling – the big male was not to be appeased. There was another sudden, vicious interlocking of bodies that left both lions bleeding from the slash and counter-slash of claws. Briefly they broke apart; but, when Scar returned to the attack, the nomad turned and tried to drag himself away. Scar did not go after him. In the very first skirmish, he had hurt the other lion badly. Twice he had bitten him in the neck. And again, in their second clash, he had caught him just above the hips, felt the crack of bone as his canines drove home. His own wounds were superficial. Blood dripped from his muzzle, but his loose coat and heavy mane – and the inexperience of the younger lion – had saved him from serious injury.

The nomad's body was found two days later. Somehow, though his spine had been bitten through, he had managed to drag himself to some thick cover before he died. Soon afterwards the three surviving Wageni nomads left the area. Scar, meanwhile, had found his lionesses lying close to a zebra kill near the airstrip murram pits and, for a while, an uneasy peace returned to the Marsh.

*   *   *

The turmoil brought about by the ambitious nomads was not confined to the Marsh pride. All through November the Miti Mbili group found themselves pestered from two sides. Their lack of pride males had not only lured the two wandering Marsh males, Brando and Mkubwa, but had also attracted the interest of the two big adult nomads whose presence near the Marsh had already troubled Scar.

Despite their attentions, however, the behaviour of the Miti Mbili lionesses remained an enigma. Not since the old days with Dark Mane and Old Man had any males succeeded in overcoming the curious reluctance of Notch and Young Girl, the two remaining adult lionesses, to accept a stable pride relationship. Males were tolerated only so long as either Notch or Young Girl was in oestrus. At all other times the two lionesses avoided them, though sometimes – albeit reluctantly and with some violence – they would give up their kills to their hungry suitors. As for the cubs and sub-adults, they were so unsettled by the strangers that the group's own young males were themselves driven out to become nomads rather than endure the constant harassment.

For a while, the rivals managed to avoid each other, and neither pair managed to win the acceptance of the Miti Mbili lionesses. At first the two nomads were cautious. They were experienced in the ways of living in alien territories, and would move off whenever they sensed the approach of Brando and Mkubwa. If there was to be a confrontation, it would be later, when they had gauged the strength of their enemies. Both of them were at least five years old. They were big and powerful, rangy lions nearing their prime; probably related, though they did not look alike. One had a dark mane with a fringe of black behind the paler mutton-chop whiskers that sprouted from his cheeks, and black elbow tufts. The other had a blond mane, a shock-headed thatch of tousled hair which added to his wild appearance. He was further distinguished by a split upper lip, the relic of a wound which had left him with a permanent, lop-sided grimace that partly exposed one yellow canine. Both lower canines were raked slightly forward, protruding almost like tusks from his black lips, which gave him a look of the utmost savagery when his mouth was open; and his ears, muzzle, cheeks and sandy-grey haunches all bore the battle-scars which attested to his experience.

These two were the biggest of all the nomads who had arrived during the migration. Their wild appearance quickly earned them the name of the *Shenzi* – the ruffians – for it was clear that, sooner or later, there would be violence between the Marsh lions and the newcomers.

The trouble, when it came, was swift and conclusive. In late November the Shenzi seized the initiative. Surprising Brando and Mkubwa at rest in the grass, they rushed at them with such ferocity that the two Marsh lions offered little more than token resistance before running for their lives. So thoroughly were they intimidated that, instead of returning to the Marsh to rejoin Scar, they left the Miti Mbili grasslands for good and hurried north past Leopard Gorge into the thornbush country outside the reserve. There was less competition from other lions here; the greater risk was that they would be branded as stock-raiders and speared by Maasai moran.

It was an ignominious end for the proud Marsh males. They were, after all, still in their prime. Had Scar been with them, had they been defending their own territory when they were attacked, the conflict might have had a different outcome. Instead, their languid lives as pride masters were over, perhaps for ever. No longer could they rely on the five Marsh lionesses to provide them with easy kills. No longer could they lie unchallenged in the reeds of Musiara, waiting while the smell of water lured the wildebeest from the plains. Defeat had humbled them, reduced them to the role of outcasts, condemned them to hunt for themselves, to follow the game, wandering without sanctuary across unknown savannas, avoiding the fury of hostile prides, scavenging through the lean seasons until old age and the hyaenas took them.

As for the two nomads, their victory over Brando and Mkubwa made no difference to their relationship with the Miti Mbili lionesses. Notch and Young Girl remained as intransigent as ever and, in the end, the Shenzi turned their backs on them and retraced their steps towards the Marsh.

*   *   *

With the return of the Shenzi nomads from Miti Mbili, the struggle for the Marsh was renewed afresh. Once more the roars of the rival males

One had a dark mane with a fringe of black behind the paler mutton-chop whiskers that sprouted from his cheeks . . .

. . . a wound which had left him with a permanent lop-sided grimace . . .

The Marsh lionesses and their cubs sought to avoid the strangers . . .

echoed and re-echoed along the dark walls of the riverine forest. Again the Marsh lionesses and their cubs sought to avoid the strangers. No longer did they sprawl with indolent ease under the Marsh figs. Now when they rested they seemed fretful and suspicious, rising at once if the nomads approached, slipping hurriedly away to lose themselves in the yellowing grasslands. The pride, for so long a close-knit and cohesive alliance under the protection of its three all-powerful males, was beginning to crumble.

The younger lions were the first to suffer. As prey became scarce, the pride no longer hunted together. With no migratory herds to sustain them, the lionesses dispersed to scour the grasslands for warthog piglets and topi calves. Increasingly the five yearling cubs found themselves abandoned for long periods. Too young to fend for themselves, not yet strong enough to scavenge on equal terms with the hyaena clans, they became thin and bony-headed. Gone were the rounded, well-fed stomachs. Their ribs protruded. Their fur lost its sleek and healthy lustre. They played less often. Instead they would lie staring with dull and listless eyes, sometimes uttering plaintive grunts as they tried to locate their mothers. Their plight was a stark and poignant demonstration of their dependence on the pride system and the protective role of the territorial males.

For the young Marsh twins the disruption was no less traumatic, but

was easier to bear as they were older than their siblings. Sired by Scar two years before, they were the first of the new generation of Marsh lions, and already they had shown their promise. Even in their first year they had been a precocious pair, eager to accompany the Talek mother on her nightly wanderings. When they were still hardly more than a year old they had demonstrated their hunting instincts in a particularly bloody manner after their mother had surprised a zebra giving birth near Governor's Camp. The lioness had spotted her as she lay contracting in the grass, and while one of the twins had pounced on the mare and seized her by the head, the other began to devour the half-born foal.

The twins were fortunate to have been born into a pride of unusually active hunters who quickly initiated them in the ways of predation. All the Marsh lionesses were skilled executioners – buffalo slayers who could dispatch lesser prey such as wildebeest and zebra with ruthless efficiency. From them the Marsh twins acquired their patience and their cunning. They learned the covert skills of the ambush, and the low crouching stalk through long grass. They learned to control their eagerness, judging the timing of their explosive rushes so as not to alert their quarry too soon; and they learned the stifling holds on throats and muzzles that most quickly subdued their victims. For some time the two young lionesses had begun to show signs of independence.

The lionesses dispersed to scour the grasslands for topi calves . . .

Often storms had threatened
as thunderheads towered in
the afternoon heat . . .

Eventually, under different circumstances, they might have been recruited into the pride as mature lionesses. Instead, the threatening behaviour of the Shenzi nomads gradually drove them from the group.

A month passed. No rain fell. The November grass rains, always fickle and uncertain, had failed almost completely, leaving the Mara drier than it had been for many years. Often storms had threatened as thunderheads towered in the afternoon heat; but the few showers that washed over the plains were barely sufficient to lay the dust. Game drifted closer to the river, leaving the plains to the topi and gazelles. Hot winds sighed in the lifeless grass. Every day fleets of flat-bottomed clouds gathered on the horizon and launched themselves in slow processions across the sky without ever seeming to obscure the sun.

The Ruppell's griffon, in his eternal lazy circlings, looked down and saw the cloud shadows spilled in chains of dark islands across the desolation of golden grass. From his lofty station thousands of feet above the plains he could see the web of dusty game trails left by the migrating herds when they had retreated to the Serengeti. Sometimes, as he turned on the thermals, he caught the sudden flash of water between the dark threads of trees along the Mara River. Nothing escaped his binocular vision. He could see the dimples of old termite mounds, the tyre-tracks of Land-Rovers, and the ashy circles which were the ghosts of trees long since burned by honey-hunters and Maasai herdsmen. Crossing Topi Plain he spotted the solitary stalking form of the cheetah Nefertari hunting for new-born calves and fawns. Her fortune as a mother had fared no better since her encounter with the Marsh lions. Since then she had lost another litter, killed by a male cheetah.

Turning again, the big vulture made a wide sweep over the Marsh, where zebra were drinking at the spring. Scatterings of game were moving through the wide meadows and clearings between the reeds and the forest; buffalo, topi, waterbuck, giraffe. High over the fleecy tops of the riverine forest he sailed, where woodsmoke spiralled from Governor's Camp. He banked and returned over the reeds and pools and buffalo wallows, and saw, farther out in the open grasslands, the tawny body of a resting lion. The presence of lions was always of interest to the carrion-eater. He tipped forward and fell closer to earth to make a tighter circle of inspection. But on this occasion there was no sign of a kill, and so the griffon drifted away over Topi Plain, using the thermals to regain his favourite cruising height, until soon he was only a speck in the blue.

If Scar noticed the vulture receding into the distance, his implacable gaze did not betray it. His eyes were half-closed against the glare, giving his heavy-headed profile an expression of utter serenity but which masked nothing but a deep weariness. The big Marsh male was tired. All night he had circled the reed-beds, calling to his lionesses, marking his diminishing boundaries and voicing his warning to the elusive nomads. He knew they were not far away. Many times during his endless patrols he would pause to scrape at the ground, and would catch their scent in the damp night air. Since Brando and Mkubwa's departure the Shenzi had virtually seized control of the southern reaches of the Marsh pride territory, consolidating themselves around the airstrip and its adjoining murram pit, whose deep pools ensure a steady flow of game. The presence of Scar alone had not been enough to deter the two nomads. Outnumbered by his enemies, he had been forced to give ground, and now spent most of his time guarding the core of his territory. Here, in the fig tree forest, in the Marsh and its grassy edges, he still felt secure enough to roar defiance. But so far, though they often came by night to importune his lionesses and steal their kills, the Shenzi males had avoided a direct confrontation.

When the New Year came, Scar was still in possession of the Marsh, but the Shenzi were becoming bolder. It was now almost two years to the day that Dark Mane had died, setting in motion the saga that led to the downfall of the Miti Mbili pride and the subsequent rise of the Marsh lions. Now the Marsh pride itself was faced with the same uncertainties. The lionesses were in disarray. The Marsh twins had broken away to hunt for themselves. The cubs were starving. Everything depended upon the outcome of the struggle between Scar and the Shenzi.

Increasingly, Scar took to lying up during the day in the long grass between Topi Plain and the Marsh. Unlike the lionesses, who sought the shade early, the big male with his thick protective mane seemed less troubled by the sun, and would not seek cover until quite late in the

. . . alerting three zebra which had been drinking from the pool . . .

morning, though his heaving flanks and harsh breath did seem to indicate a measure of discomfort. From time to time his tail thumped restlessly on the ground; but whether this was due to the irritation of biting flies, or to the nagging presence of the nomads, it was impossible to say.

All day the sun had shone. This was the African winter, a time of clear skies and fine hot weather. Now, as the heat leached from the plains, Scar stretched and yawned, a cavernous yawn which displayed to the full his enormous yellow canines. He stared around him. Everywhere, game was on the move again after the midday torpor. Impala emerged from the shade of trees. Zebra which had stood with twitching ears and swishing tails now frisked and squealed in the cooling air. Gazelles pricked the skyline with their sharp horned heads as they awaited the coming of the predators.

Scar, too, rose to his feet, fully awake. One hour to sunset, and the westering light dissolving in a lambent golden glow. Soon his lionesses would also be hunting. Slowly he walked down to the spring, a dark shaggy shape in a lion-coloured landscape of swaying grassheads and lengthening shadows. As he approached the water a reedbuck exploded almost from beneath his feet, alerting three zebra which had been drinking from the pool. Somewhere by the river an elephant screamed.

For several minutes the big cat lapped steadily at the clear spring water. When he had finished, he padded into the crushed and trampled wreckage of the old year's reeds. Close by, the previous day, the Marsh twins had managed to pull down a buffalo. The victim was a cow, and very sick, and they had killed her so clumsily that tourists who had witnessed the incident were too distressed to remain and hurriedly drove away. But it was still a remarkable achievement for the two young lionesses to have dispatched such a strong animal, even though they were afterwards chased from their kill by the Murram Pit hyaenas.

In turn, the hyaenas were dispossessed by Scar. For days he had eaten little. When he was really hungry he could consume up to eighty pounds of meat at one sitting. Now he was famished. He sniffed with interest at the abandoned buffalo and thrust his head inside the cavern of ribs. He stood for some minutes with his head buried deep inside the carcass. When at last he withdrew, he looked up and saw a lion watching him far out in the grass. It was the blond-maned Shenzi nomad.

How long he had been there Scar did not know, but he saw the nomad stiffen as he turned to face him. Scar began to roar, lifting his broad muzzle to hurl the full strength of his challenge in the face of his enemy. The sun

The Marsh twins had managed to pull down a buffalo . . .

sank behind him in a crimson blaze and the Marsh pools turned blood-red in its dying rays. As he watched, the nomad began to strut towards him. Then he, too, began to roar, his voice ugly with menace. A hundred yards from the reeds, the strut became a trot, and suddenly he was coming in a rush.

For a brief moment, as they came together, the two lions were up on their hind legs like boxers, slashing and ripping with heavy paws. Then the blond lion went down under Scar's onslaught and the two of them rolled across the ground, fighting furiously. Suddenly the nomad broke away, bleeding heavily about the chest where Scar's claws had hooked out great tufts of mane.

All this time, unbeknown to Scar, the dark-maned nomad had been watching from the long grass a little further off. Now, as the two combatants stood back and glared, gasping for breath through snarling teeth, he seized his chance. Slinking in from behind, he hurled himself at Scar's haunches. Again the Marsh echoed to their hideous growling as all three lions locked and grappled, each striving to avoid the disabling bites of the other as they raked and tore with their claws. What had begun as a ritual dispute over territorial supremacy had developed into a murderous brawl. Scar was soon bleeding from a dozen wounds. In the heat of battle he felt no pain, but he sensed the hopelessness of the struggle. In the end it was not resolved through any single decisive bite. When life itself was threatened, self-preservation became the only instinct. Summoning all his remaining strength, Scar shook himself free and fled towards the river.

Tree-frogs were chiming in the dusk as he walked along the wooded banks, looking for a place to cross. All aggression had drained away. For the first time he felt his wounds and lowered his head to lick a throbbing paw. He could not stay. In running from the Shenzi nomads he had lost his pride and his territory. Once again he was a lion alone, a fugitive. But where could he go? Not to the south. There, the Paradise pride was strong. To the north roamed the Gorge pride. The east, too, was hostile. That way lay nothing but drought and, beyond, the powerful Talek prides. Only the west held out any hope. He would take his chances in the unknown country of the Mara Triangle.

As he walked down the bank he paused, as if reluctant to leave his old hunting grounds. This time, when the wildebeest returned after the long rains, he would not be waiting for them in the Marsh. But his line would continue in the blood of his daughters. In time they would raise cubs of their own, and teach them to hunt on Topi Plain, and in the reed-beds of Musiara. Wearily he raised his torn muzzle. For the last time his roar thundered across the Marsh. A mile away at Governor's Camp, the Africans heard him and nodded knowingly. They understood what he was saying. 'Whose land is this? Whose land is this? It is mine; it is mine.' By the time the last rumbling echo had died away, he had already crossed to the other side and disappeared into the night.

February 1980-September 1981

Epilogue

When we first set out to portray the lives of the Marsh lions it was our intention to close their story with the account of Scar's fight with the Shenzi nomads. His subsequent retreat across the Mara River provided a fitting finale, just as the defeat of Old Man two years earlier had provided their story with a natural beginning. But, although Scar's overthrow ended the reign of the Marsh pride, it was not the end of the lions themselves. In nature, there is no end and no beginning. In time, the Marsh lionesses accepted the victorious Shenzi and a new dynasty of Marsh lions was formed.

Thus, having followed them so far – and the other predators whose lives they shared – we felt unable to leave them. What, we wondered, would become of the wall-eyed leopard, the cheetah Nefertari and the wild dogs of Aitong? And what of the Marsh twins, of Notch and Young Girl on Miti Mbili Plain, and Scar across the river? By good fortune we were able to watch them for another two years, during which time the story of the Marsh lions took an unexpected twist and an old friend reappeared.

*   *   *

SCAR WAS not the only familiar character to fade from the Marsh and its surroundings. Over the next two years, many of the animals we had come to recognize as individuals while following the Marsh lions were to disappear. In human terms their lives seemed pitifully short. Their years were measured by the steady pulse of the seasons, by the coming of the rains and the marching wildebeest. Seldom did lion, leopard or cheetah witness more than a dozen migrations in a lifetime. Yet, while they lived, their wild presence carried with it a constant sense of imminent drama. And their deaths, so often unseen and unrecorded, were all the more poignant because of the mystery which shrouded the manner of their going.

Brando was the first to vanish. Like Scar, he was still in his prime when he was chased away by the Shenzi nomads. Early in 1980 he turned up again briefly just outside the reserve near the Mara River Camp, still accompanied by his fellow exile, Mkubwa. Both males appeared to be interested in a small group of lionesses who were resting close by with their half-grown cubs. This was to be Brando's last appearance. Since then, the handsome male whose fight with Old Man had paved the way for the creation of the Marsh pride has not been seen, and we do not know what has happened to him.

Mkubwa was luckier. By now he, too, had disappeared, and we feared that he may have met with a sudden and violent end. But Mkubwa, the biggest of the three Marsh males, was ever the opportunist. Unknown to us, after failing to establish a permanent liaison with the lionesses near the Mara River camp, he eventually crossed the river into the Triangle. There he managed to join the Kichwa Tembo pride, whose hunting grounds stretched along the foot of the escarpment as far as the Sabaringo Lugga. At the time, however, we knew none of this, and did not meet up with Mkubwa again until the following year.

They emerged, blinking in bewilderment at the unfamiliar daylight . . .

Meanwhile, another well-known resident had vanished from the scene. This was the big leopard, the wary, wall-eyed male who had haunted Leopard Gorge for so many years. Early in the morning of November 23, 1979, four sub-adult lions from the Gorge pride managed to corner a sick buffalo cow among the rocks of Leopard Gorge. One of the adult lionesses from the pride was also nearby, watching from a ledge. It was a messy kill. The youngsters were clumsy and inexperienced and, in spite of her sickness, the buffalo defended herself so fiercely that it was nearly noon before she died.

So intent were we on the drama being played out before us that we had failed to notice another watcher concealed in a shady cleft on the opposite wall of the gorge. Only when the pride settled down to feed did we see the leopard. He must have been there all the time, quietly watching both us and the lions with his one good eye.

The New Year came, but there were no further reports of the wall-eyed leopard. Maybe the poachers had caught up with him at last. We may never know; but it is somehow fitting that his disappearance, like so much else about him, should be obscure and mysterious.

By late 1980, the young female leopard sired three years earlier by the old wall-eyed tom was heavily pregnant and about to produce her first litter. Having inherited all her father's caution, she was extremely shy and seldom ventured forth except under cover of darkness.

Some time towards the end of November 1980 she gave birth to two cubs in a cave among the rocks of Leopard Gorge. For the first few weeks the twins remained hidden inside the cave. Then, one morning in January 1981, they emerged, blinking in bewilderment at the unfamiliar daylight, and sat on a rocky ledge at the entrance to the cave. Their eyes were still a

. . . two young lions prowling among the rocks . . .

kittenish blue and their spotted fur the soft grey of thistledown, merging so perfectly with the mottled rock that it was hard to pick them out unless they moved.

Not long afterwards Jonathan returned to the gorge, hoping to watch the mother leopard and her cubs. Instead he found one of the Gorge pride lionesses and two young lions prowling among the rocks. They were grimacing as they sniffed along the ledge, and it was clear from their behaviour that they had caught the scent of leopard. Had they already found and killed the cubs? For three months there was no sign either of the cubs or of their mother. But then, in mid-May, the twins were seen resting together in a tree near Leopard Lugga. Somehow they had managed to evade the Gorge pride and escape with their mother.

Their luck did not last. Shortly after this latest sighting, a driver from Governor's Camp saw the mother leopard calling to her cubs from a tree. Eventually one of the twins emerged from a thicket looking very angry and frightened. Moments later, a lioness appeared from the other side of the bushes with fresh blood on her face. The mother leopard continued to call from her tree, but the second cub never reappeared.

In September 1981 a large male leopard was seen several times along the airstrip lugga, while a young female which established her home range on Rhino Ridge in the same year has become so habituated to the presence of tourist vehicles that we were even able to observe her stalking impala in daylight. Her daytime appearances are an encouraging

indication that, after many years of poaching, leopards are at last increasing again in the Mara.

The cheetahs are also thriving, especially on the plains just outside the reserve where there is less competition from lions and hyaenas. Nefertari is still alive but she has remained a tragic figure, having lost at least two more litters. The last was in March 1981 when she produced four cubs near Airstrip Lugga. Within a week, one had died from natural causes. Two days later, plagued by tourist vehicles and alarmed by the sudden arrival, some two hundred yards away, of the Marsh lions, she moved the three survivors to a new hiding place in a patch of long grass. There she left them while she hunted. That night, the long rains began in earnest. Exposed in the open, the tiny cubs had little chance of survival. Next morning, only one remained alive. It lingered on for a few more days until, lacking the collective warmth of its siblings when its mother was away, it, too, perished. In all the time we had known her, Nefertari had managed to raise only two sons and a daughter.

Fortunately, other cheetahs have fared better. Towards the end of the 1980 migration, a female produced a litter of cubs on Paradise Plain. The whole family were still together four months later, when they left the reserve and settled for a while in the thorny scrublands near the Mara Buffalo Camp. But some time during the rains, two of them disappeared. When the 1981 migration arrived, the survivors returned to overlap Nefertari's home range. By now the cubs were almost as big as

After many years of poaching, leopards are at last increasing in the Mara . . .

A female produced a litter of cubs on Paradise Plain . . .

their mother. Already they had learned how to catch young Thomson's gazelles; and in September we watched them flush a reedbuck from a patch of long grass at the foot of Rhino Ridge. All four cats were after it in a flash, and the mother caught it after a short chase. They were very hungry, and very nervous, lifting red muzzles to scan the plains between each bolted mouthful. There was no cover, either from the mid-morning sun or from the gaze of predators. Their bellies were thin, and it was clear from their behaviour that they must have been losing many kills to lions and hyaenas. That is the price all cheetahs must pay when more powerful predators are common. But at least the cubs had shown they could survive. Every day they were becoming less vulnerable. Soon they would be old enough to fend for themselves, and would drift away from the land of the Marsh lions to lead their own lives. But what would happen to their mother? Would she, too, be forced to seek new hunting grounds to avoid the robber prides? Or would her life mirror that of the ageing Nefertari? One thing is certain. If she does choose to remain near the Marsh, she will be inheriting all the pressures which Nefertari has endured for so long.

*   *   *

The wild dogs, too, are finding it increasingly hard to survive among the all-powerful prides and hyaena clans. When the Aitong pack returned to their old denning area in 1980, one of the five pups they had raised the previous year was missing.

As usual, the dogs remained throughout the migration, hunting across the Plain of Stones, and sometimes making longer forays into the Miti Mbili and Marsh pride territories. On September 2 1980 the pack was found resting under a tree on the far side of Miti Mbili Plain, and it was clear that three of the previous year's pups, all females, were sick. Their bodies were thin. They gazed into the distance with lacklustre eyes, and from time to time their hindquarters trembled uncontrollably. Two days later, the entire pack had disappeared and could not be found. When the dogs reappeared again on September 5, only one of their offspring was still with them. This was the young male, and it seemed likely that the three females had died.

A week later the dominant male, Black Dog, himself narrowly escaped death when he was caught and pinned down by a lioness on the Plain of Stones. Luckily the rest of the pack came to his rescue, harassing the lioness so fiercely that she was forced to let him go.

The pack was not seen again until November 1980. As they appeared, running along the skyline towards Aitong, we saw to our dismay that the young sub-adult male, the sole survivor of the previous year's pups, was not there. The pack was once again reduced to three animals: Black Dog, the Bitch and the brindled male.

The dogs stayed near Aitong for some time. Often they were seen digging near their old dens, but they did not breed that year, and in February 1981 they came south through the Marsh lion country, where they lingered briefly on the plains around the Governor's Camp airstrip.

There, one evening shortly before dark, they killed a young wildebeest which had failed to return to the Serengeti. It was still twitching in the

Already they had learned how to catch young Thomson's gazelles . . .

One of the five pups they had raised the previous year was missing . . .

grass as one of the Marsh twins came trotting towards them. At once Black Dog and the Bitch ran off, voicing their displeasure, but the brindled dog remained and seemed determined to try and snatch a last few mouthfuls, for he looked even thinner and hungrier than his companions. Unfortunately, he failed to see the second Marsh twin stealing through the grass behind him. She grabbed him in her jaws, where he wriggled and yelled helplessly. Luckily for the dog, a French photographer had been watching the entire drama unfold. As soon as the Marsh twin caught the dog, he drove forward, forcing the lioness to release her victim, who ran off seemingly unhurt.

But fate had not finished with the Aitong pack. When next they appeared in mid-July 1981, at the beginning of the migration, the brindled male was no longer with them. Perhaps he had again been caught by a lion – this time with no human observer to come to his rescue.

Now only two dogs were left: Black Dog and the Bitch; and in September we witnessed two incidents which portrayed with brutal clarity the increasing hopelessness of their plight. One morning we were out on the plains at sun-up, scanning the slopes of Rhino Ridge through binoculars. A sudden scattering of wildebeest caught our attention. At first we thought they were being pursued by hyaenas, but as the chase came more sharply into focus, we recognized Black Dog and the Bitch. Swiftly they singled out a wildebeest calf and pulled it down after a brief struggle. But even as we watched, half a dozen hyaenas appeared as if from nowhere to converge on the dogs and drive them from their kill. Outnumbered three to one, the dogs had no choice but to retreat.

The next time we saw the dogs, the wildebeest had begun their return trek to the Serengeti and were streaming out of the reserve in endless columns. To reach the river, the wildebeest had to run the gauntlet of the Musiara grasslands, and the Marsh lions took a heavy toll. The Marsh twins, in particular, had been unusually active, having positioned themselves in the very path of the herds as they skirted Airstrip Lugga.

Waiting to see if they would kill again, we had stopped our vehicle near the murram pits where the twins had set their ambush. Dusk was falling, the light was fading and coqui francolin were beginning to call from the grass around us. Suddenly, about a quarter of a mile away, the plodding columns of wildebeest broke and scattered like crows in a gale. And there were the dogs, snapping at their heels as they tried to single out a calf.

It took the two dogs a long time to run down their kill. In the end it was the Bitch who finally managed to grab the calf by its tail, and Black Dog was panting with exhaustion by the time he arrived to assist her. But all to no avail. We had not been the only witnesses: as soon as they heard the calf bellowing, the Marsh twins had raised their heads to watch. Now, as the dogs tried to subdue the struggling wildebeest, the two lionesses padded towards them.

Once again the dogs were robbed of their kill by predators more powerful than themselves. They leapt away in fright, then turned and stood a little way off, their thin bodies shaking as they barked in

We realized for the first time
just how very old he was . . .

frustration, until one of the twins rose and drove them away.

We saw them one more time in September, asleep under a solitary acacia far out on Miti Mbili Plain. It was mid-afternoon and they were lying together, the Bitch with her chin resting affectionately across Black Dog's hairless haunches. Both were panting in the heat. We were very close; and we noticed Black Dog's canines. They were no longer sharp and white, but had been worn away to brown stumps. We realized for the first time just how very old he was.

We left them, still sleeping, with a deep sense of sadness. It is impossible to believe that we may never again see their lean forms racing at dawn along the red horizons. But, unless they manage to recruit new members from the Serengeti or the Loita Plains, the wild dogs of Aitong seem doomed.

*   *   *

Tragic, too, is the fate of the rhinos, poached to the edge of extinction as the demand for horn continues. At the beginning of 1981, despite vigorous efforts by the Kenyan Government, only just over a dozen rhinos remained in the Mara, where a decade ago there had been at least ten times that number. By the end of the year, at least one more had been shot, leaving an orphaned calf, and another had disappeared. In the Marsh lion country and in the territories of the neighbouring prides, we know of only two surviving rhinos. One is a young adult male from the Paradise area; the other is the young female whose mother had herself been killed by poachers, and whose harassment by the young Miti Mbili lions is described in chapter ten.

She was seen regularly in the company of the male rhino . . .

By the end of 1980 the female had become old enough to mate, and was seen regularly in the company of the male rhino until March 1981, after which they went their separate ways again. In July the female was shot by poachers and hit in the foreleg. The park staff managed to immobilize her with a tranquilizing dart and to remove the bullet from her leg. Today she is still alive, a lonely figure in the emptiness of Miti Mbili Plain. But it can only be a matter of time before the poachers try again. For sixty million years her kind have wandered across this vast continent. Now their future is bleak, and the Mara will be the poorer if they are allowed to disappear completely.

*   *   *

Fortunately no such threats face the lions. For the time being at least, their future seems secure. Within the protection of the reserve the prides are now so numerous that, mile for mile, the Mara probably contains as many lions as anywhere in Africa. The wildebeest likewise continue to multiply, providing an endless source of food, and the last two years have seen the most spectacular mass migrations in living memory. To that extent the Maasai country has not changed. The essential drama of its wild grasslands remains as it always was, the story of the lion and the wildebeest.

But what of the Marsh lions themselves? As usual the struggle for supremacy between the rival males brought about a period of intense hardship and disruption for the lionesses and their cubs; and its repercussions spread far beyond the confines of the Marsh pride boundaries. Having successfully displaced Scar, Brando and Mkubwa, the Shenzi nomads tried once more to establish themselves with the Marsh lionesses, but none of the five adult females was yet ready to accept them. There were still five dependent youngsters to protect, as well as the young Marsh twins, and they all went in fear of the strange new males. Sometimes even the sight of the pair of them approaching in the distance was enough to put the entire pride to flight.

The pride remained in disarray throughout the first half of 1980. Concern for the cubs kept the lionesses constantly on the move. In vain they tried to avoid the Shenzi, who dogged their footsteps, stole their kills and marked the Marsh with scraping paws to claim it as their own.

With prey less plentiful – the migration would not arrive until July – the lionesses were forced to travel farther afield to hunt warthog and topi.

The Mara will be the poorer if they are allowed to disappear completely . . .

And all the time the cubs grew thinner, separated from their mothers and chased from kills by the hostile males. Eventually, when the long rains broke, the ragged pride forsook the Marsh altogether and moved away towards Miti Mbili. For a short time, the Shenzi stayed behind, as if asserting their right to the territory they had won, until the rising floodwaters and their own hunger compelled them to go in search of the departed females.

With the Mara River in spate and every lugga overflowing, it became increasingly difficult to follow the lions until the rainy season had passed. Even then, any vehicle straying off the regular game-viewing trails could suddenly find itself sinking up to its axles in the notoriously soft black cotton soil. All through April the pride remained virtually inaccessible in the waterlogged fastness of the western plains. At the beginning of May, however, it soon became apparent that the situation had changed dramatically. All the youngsters had gone, leaving the Shenzi in full possession of the five adult lionesses, who had at last accepted the two males as their mates.

The Marsh lions had become a true pride again, but at what cost? Of the five surviving cubs born to the Marsh sisters in 1978, four had vanished without trace during the rains. The fifth youngster, a female, had survived, though not entirely unscathed. The bold black tassel at the tip of her tail had been bitten off in a brawl with the nomads, earning her the name of Stump.

Soon afterwards she must have broken away from the pride and joined the Marsh twins. These were Scar's daughters, the first cubs to be born to the old Marsh pride, and a few months older than she was. They had left the pride earlier, forced to fend for themselves rather than endure the unbridled antagonism of the Shenzi.

Together the three lionesses had formed a small splinter group which hunted on the grasslands around the Governor's Camp airstrip. At first they were still greeted amicably enough by the adult lionesses of the new Marsh pride, encamped to the east on Miti Mbili Lugga. But, as the year progressed, it became increasingly obvious that the break was irrevocable. Even without the disruptive presence of the Shenzi males it is quite likely that the trio would still have gone their separate ways. With five healthy adult females in the existing pride, there was no room for the three extra lionesses. So Stump and the Marsh twins became exiles, rather as their own mothers had been at the beginning of 1978, before they mated with Brando, Scar and Mkubwa and formed the original Marsh pride.

By the end of January 1981 the new Marsh pride was fourteen strong. The barren Talek lioness had disappeared in late 1980; but the four remaining lionesses all produced new litters. The Talek mother was the first to give birth. Her cubs were born in August, in a thicket beside the Miti Mbili lugga. By October, two of the Marsh sisters had also produced new families, and the third sister gave birth early in the New Year. In their first year as resident pride males, the Shenzi had managed to sire a dozen cubs, of which eight were still thriving a year later.

Meanwhile, the Miti Mbili lions had also been undergoing a period of dramatic change. After nearly three years without the permanent protection of resident pride males, Notch and Young Girl had finally accepted a pair of nomads who had appeared in June 1980, not long before the arrival of the wildebeest.

In their previous encounters with amorous males, the Miti Mbili lionesses had always reacted with hostility, fearing for the safety of their cubs. Now the three eldest cubs were themselves adult lionesses. The rest had disappeared during the upheaval which followed the collapse of the old Marsh pride, when the Marsh lionesses invaded the hunting grounds of the Miti Mbili group in their desperation to escape from the Shenzi. The result had been doubly catastrophic for the Miti Mbili youngsters, who found themselves attacked not only by the Marsh lionesses but also by the two formidable Shenzi males who had followed the females in search of food.

Initially, the Miti Mbili group had retreated still farther east, but resistance from the Olare Orok prides had forced them towards Rhino

The four remaining lionesses all produced new litters . . .

Notch had finally accepted a pair of nomads . . .

*In October 1980, one of
Notch's daughters produced
cubs . . .*

Ridge. Their new home was less favourable than the land they had
relinquished to the Marsh lions; inferior, too, to the rich grasslands and
sheltering thickets of the Paradise pride to the south. But here at least they
were secure, and were able to move out at night to hunt on the plains
below. It seemed strangely ironic that the Miti Mbili group should now
find themselves on Rhino Ridge; for that was where Scar, Brando and
Mkubwa had lived at the time when Notch and Young Girl had been part
of Old Man's pride. The fortunes of the two warring prides had come full
circle; but, if this point represented the luck of the Miti Mbili group at its
lowest ebb, the arrival of their new resident males brought new hope and
new life. In October 1980, Young Girl and one of Notch's daughters each
produced cubs. At last their troubles seemed to be at an end. With two
vigorous males, five adult lionesses and five healthy young cubs, the pride
was stronger than it had been for years.

All this time the Marsh itself remained unoccupied. When the
wildebeest returned in 1980 we expected to see the new Marsh pride with
their Shenzi males drawn back to the reed-beds by the abundance of prey

in the area. But, although the lionesses occasionally hunted along the grassy margins, they seemed unwilling to reclaim their old haunts, preferring instead to roam on Topi Plain, where they ambushed the game from the surrounding luggas.

In August that year, while the migration was at its peak, we made an extraordinary discovery. One of the drivers reported that he had seen three big male lions together in the park-like meadows between the Marsh and the forest north of Governor's Camp. Furthermore, he said, one of them was a battle-scarred old lion with only one eye.

Who could these strangers be, we wondered? The only lion we had known who could fit that description was Old Man, but we had never seen him after his battle with Brando and had always assumed that he had died. Besides, who could his two companions be? Probably the trio which the driver had seen were nomadic males who had drifted in with the migrating herds from another part of the reserve. But, when eventually we found them resting beneath a tree not far from the spot where they had first been seen, we received a double surprise. The one-eyed male was indeed Old Man – and his two companions turned out to be none other than the missing Marsh males, Scar and Mkubwa. They were both in excellent condition, older and greyer but well-fed and alert. It seemed strange to see them consorting with Old Man, for in the old days they had been his rivals, and were instrumental in bringing about his downfall. As for Old Man himself, he was more grizzled than ever. One of his canines had been broken, and he had completely lost his wounded left eye. Yet he, too, seemed sleek and fat, and his mane was still magnificent; he was clearly an animal with remarkable powers of survival.

Later, having talked to some of the drivers from the Kichwa Tembo tented camp on the opposite side of the Mara River, we were able to piece together Old Man's history after he had been cast out of his pride. They told us that in 1978 an old male had appeared in the area. His face had been badly clawed and his left eye injured. The wound had never healed, and the eye itself was later gouged out in another brawl. The dates and description fitted perfectly. There could be no doubt that it was Old Man.

After crossing the river, he began to trail the lionesses of the local Kichwa Tembo pride. On several occasions they were seen fighting over kills but, fortunately for the old exile, there seemed to be no resident males to chase him away. The only rivals in the area were a pair of sub-adult males which the more experienced veteran had been able to drive off.

By the end of the year the females had come to accept him. Once again he had lionesses who would provide him with kills and, despite his age, he mated vigorously with all four of them. Once more he was a pride male: master of the Kichwa Tembo pride.

He was still there a year later when the Marsh pride broke apart and Scar and Mkubwa both crossed the river to escape from the Shenzi. Within a few months they, too, had joined the Kichwa Tembo pride. There were no young cubs, so the lionesses accepted them, and Old Man

The one-eyed male was indeed Old Man . . .

Despite his age, he mated
vigorously with them . . .

offered no opposition. Together, Scar and Mkubwa were easily able to dominate him; but, perhaps because they did not regard him as a dangerous rival, they had not driven him away.

With three males in attendance, the pride thrived. The lionesses felt secure and by the end of 1980 Scar and Mkubwa had sired four new cubs. The Kichwa Tembo territory is ideal for the lions in every respect except one. Often, during the migration, the herds move out and cross the river in search of fresh grazing, leaving the pride with little game to hunt. Then the temptation to follow the wildebeest becomes too much, especially when they can be seen and heard in huge numbers on the opposite bank.

Such was the case in August 1980 when, finding the Marsh unoccupied, Old Man, Scar and Mkubwa returned to their former haunts. That year, and again in 1981, from July onwards when the water was low, they crossed and recrossed many times. Sometimes their lionesses came too; but, so far, they have made no attempt to lay fresh claims to the Marsh: they always return to Kichwa Tembo.

The Kichwa Tembo territory is ideal for the lions . . .

One question still puzzled us. Where did the lions cross the river? Although there were several shallow rocky stretches where it would be easy for a lion to wade, we had never actually seen them crossing. However, in September 1981, we surprised a young male lion drinking at the foot of a steep bank a few hundred yards upstream from one of the main crossing places used by the wildebeest. The lion was a stranger, one of four nomads which had arrived recently in the Paradise pride territory. All of them were nervous, not only of the Paradise males, but also of vehicles. Our sudden arrival at the top of the bank startled the nomad; but instead of running he plunged into the river and swam powerfully and easily to the other side, where he jumped out, shook his bedraggled mane, and then bounded away into the thickets. Our curiosity about the crossings was satisfied. We had seen that the river, even in its deepest stretches, was no obstacle to a determined lion.

The presence of this particular band of nomads caused some disruption within the Paradise pride, which is at present without very young cubs. But elsewhere, then and during the previous year, there was no great influx of nomadic lions, and both the new Marsh pride and the resurgent Miti Mbili pride on Rhino Ridge were able to enjoy a period of relative stability. For a time, during the 1981 migration, a pair of young nomads did appear near the airstrip and stayed long enough to mate with Stump and one of the Marsh twins. But they were soon chased off by the Shenzi males, who frightened them so badly that one was forced to dive down a warthog burrow to escape a mauling.

Up on Rhino Ridge, the Miti Mbili pride continues to thrive. Young Girl is no longer young, but she is still a beautiful lioness, and the old virago, Notch, is a mother again. She was seen with a single cub early in 1981, but continued motherhood does not seem to have mellowed her unpredictable nature.

One day in January we watched one of her adult daughters kill a warthog on the plains below the ridge. The pig was a large adult boar with big tusks. When it realised it could not reach the safety of its burrow, it turned and confronted its pursuer. The lioness stopped in her tracks and there now began an extraordinary duel as the warthog lunged forward every time she moved to try and outflank him. Unfortunately for the pig, however, Young Girl appeared and caught him from behind. He sank on his haunches, then went under altogether as the two cats drowned his squeals with their own hungry growling.

Hardly had the pig ceased to twitch when Notch, a third lioness and the small cubs came trotting down from the ridge, and soon they were all lying round the carcass. As they began to feed, they were joined by the smaller of the two pride males, who snatched away the entire pig's head and stood with it in his jaws. The lionesses and cubs stared at him nervously, but sensed that he was in a mood to tolerate them and after he had begun to gnaw at the carcass they, too, resumed their meal.

All this time the second male had remained aloof, lying on the grass a short distance away. Suddenly, without warning, he rose and hurled

himself, growling furiously, into the midst of his feeding family. The speed of his rush caught them all by surprise. Young Girl raised her bloody muzzle and snarled in fright; but he bounded past her, intent on attacking the smaller male who had leapt sideways to avoid him, nearly flattening two of the cubs. There was a brief flurry of blows; then the weaker male rolled over to signal his submission. The aggressor did not return to the kill. Instead he moved off and lay down in the shade. Such quarrels are common when there is little to eat except warthogs. Sometimes, too, the males become more irritable when mating is imminent. On this occasion no damage was done; but it demonstrated the frightening power of the males, and also the extreme vulnerability of small cubs when sudden fights break out.

The new Marsh pride are still encamped on Miti Mbili lugga, where they have remained since the migration, preferring to ambush their prey at night and then lie up by day in the thickets. The Shenzi are still with them in almost constant attendance. Yet curiously they have not returned to the Marsh. Having pushed the old Miti Mbili group back towards Rhino Ridge, the lionesses seem content to remain in the vicinity of Miti Mbili lugga. With young cubs to protect, they may be nervous of venturing too far west towards the reed-beds of Musiara for fear of meeting the Kichwa Tembo trio, who are still regularly crossing the river

They sensed that he was in a mood to tolerate them . . .

He rose and hurled himself, growling furiously, into the midst of his feeding family . . .

to prowl around the marshland margins.

In September, we had pitched our tents in a beautiful stretch of riverine forest near a spot known as Crocodile Camp. Every night and again at dawn we heard lions roaring all around us. On one particular evening the sound was so close it made the air vibrate, and almost immediately one of our African camp staff rushed into the tent to tell us that a lion had come into camp. We grabbed torches and ran past the campfire to see a big male lion standing no more than twenty paces away. Even by torchlight we could recognise him by his scarred right nostril. It was the old Marsh male, Mkubwa. For a second or so he stared back at us, pale eyes shining. Then he gave a startled *whoof* and bounded away to rejoin Scar and Old Man, who had been standing at the edge of the forest. Our cook had wisely taken refuge in a tree, yet we ourselves, rashly perhaps, had felt no fear. We were sure that Mkubwa was simply being curious and had meant us no harm. It was, after all, his land as much as ours.

Now another year is ending. The wildebeest have returned to the Serengeti, pouring across the Mara River in such a press of bodies that at

one point we counted the corpses of the drowned and trampled floating past us at the rate of twenty-five a minute. Again the plains are spent and empty. For the lions, the time of hunger has returned.

In September we had watched Stump and the Marsh twins feasting on a glut of kills as day after day they ambushed the retreating columns of wildebeest. Scar's daughters were now expert predators, stalking their prey with consummate skill. When they killed, wrapping their huge furry paws round their victims as they held them by the throat, the end was swift and clean; more like an act of love than the ending of a life. It was a far cry from their sub-adult days, when we had watched them attack a zebra and clumsily trap it in a pool on Leopard Lugga, only to lose it to the Murram Pit hyaenas.

They had survived those first perilous months of enforced independence, and the bond of affection which seemed to bind them was as strong as ever. Often, even though they were almost fully grown, they would play together, romping and splashing through the pools on Topi Plain in the aftermath of the rains. Now, with their companion Stump, they have

The new Marsh pride are still encamped on Miti Mbili Lugga . . .

become the potential nucleus of a new Marsh pride.

For five years we have been privileged to follow the Marsh lions, to become familiar with them as individuals, and to share their pristine world on the African plains. Now, although we must leave them, their story continues. Perhaps Scar and his companions will leave the Kichwa Tembo lionesses and form a new pride with Stump and the Marsh twins. Maybe the Shenzi will return. Or perhaps the small cubs we saw playing with Notch and Young Girl on Rhino Ridge are destined to become tomorrow's Marsh lions. Meanwhile, Musiara Marsh lies empty. No lions have held it since the Shenzi moved out with the old Marsh lionesses more than two summers ago. But nature abhors a vacuum. Sooner or later, a new generation of lions will arise. Once more its dense cover will hide their crouching tawny shapes as the wildebeest plod to their death in the tall reeds, enacting an endless story as old as Africa itself.

We had watched them attack a zebra and clumsily trap it in a pool . . .

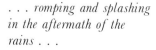

. . . romping and splashing in the aftermath of the rains . . .

OPPOSITE: *Sooner or later, a new generation of lions will arise . . .*

Family Tree of the
Marsh and Miti Mbili Prides

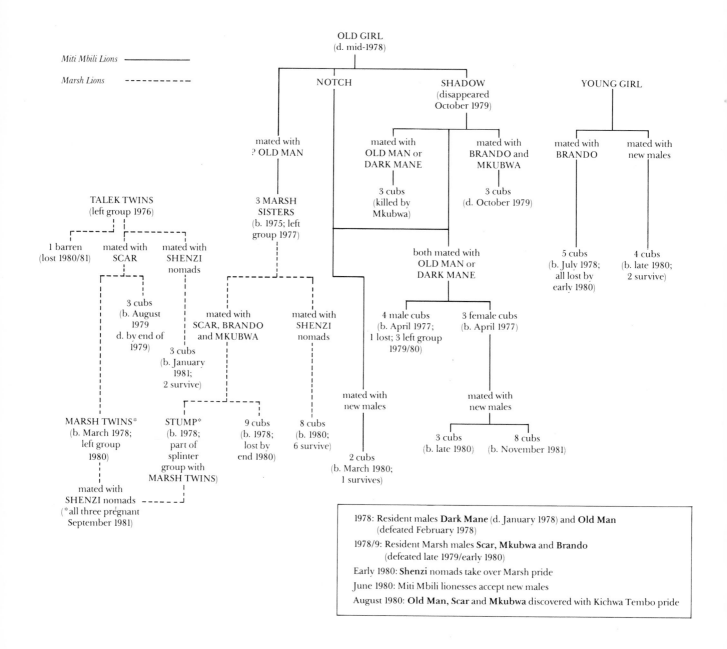

OLD GIRL
(d. mid-1978)

Miti Mbili Lions ————————

Marsh Lions — — — — — —

NOTCH

SHADOW
(disappeared
October 1979)

YOUNG GIRL

mated with
? OLD MAN

mated with
OLD MAN or
DARK MANE

mated with
BRANDO and
MKUBWA

mated with
BRANDO

mated with
new males

3 MARSH
SISTERS
(b. 1975; left
group 1977)

3 cubs
(killed by
Mkubwa)

3 cubs
(d. October 1979)

5 cubs
(b. July 1978;
all lost by
early 1980)

4 cubs
(b. late 1980;
2 survive)

TALEK TWINS
(left group 1976)

1 barren
(lost 1980/81)

mated with
SCAR

mated with
SHENZI
nomads

both mated with
OLD MAN or
DARK MANE

3 cubs
(b. August
1979
d. by end of
1979)

mated with
SCAR, BRANDO
and MKUBWA

mated with
SHENZI
nomads

4 male cubs
(b. April 1977;
1 lost; 3 left group
1979/80)

3 female cubs
(b. April 1977)

3 cubs
(b. January
1981;
2 survive)

mated with
new males

mated with
new males

MARSH TWINS*
(b. March 1978;
left group
1980)

STUMP*
(b. 1978;
part of
splinter
group with
MARSH TWINS)

9 cubs
(b. 1978;
lost by
end 1980)

8 cubs
(b. 1980;
6 survive)

3 cubs
(b. late 1980)

8 cubs
(b. November 1981)

mated with
SHENZI nomads
(*all three pregnant
September 1981)

2 cubs
(b. March 1980;
1 survives)

1978: Resident males **Dark Mane** (d. January 1978) and **Old Man**
(defeated February 1978)

1978/9: Resident Marsh males **Scar**, **Mkubwa** and **Brando**
(defeated late 1979/early 1980)

Early 1980: **Shenzi** nomads take over Marsh pride

June 1980: Miti Mbili lionesses accept new males

August 1980: **Old Man, Scar** and **Mkubwa** discovered with Kichwa Tembo pride

Author's Note

THIS BOOK is a joint effort which attempts to do justice, in words and pictures, to Kenya's finest game reserve. It is the result of a chance meeting between Jonathan Scott and myself at the Mara River Camp in 1978. It was my fifth visit to the Maasai Mara, but the first time that I had been shown the pride of lions which then occupied the area known as Musiara Marsh. With Jonathan as my guide I was introduced to some of the pride members who were later to become so familiar to me: the big resident males, Scar, Brando and Mkubwa; and the three handsome lionesses known as the Marsh sisters.

Jonathan, I discovered, was not only a first-class field naturalist, but also a talented wildlife artist and photographer. He was keen to have his work published. I had long wanted to write a book about Kenya and its animals. How could we collaborate?

The answer came one evening by the camp fire on the banks of the Mara River. Ever since his arrival in the Mara, Jonathan had kept a detailed account of each day's game-viewing, recording the animals he had seen and setting down his observations of their daily lives. Inevitably, lions figured prominently, and one group in particular: the pride I had seen at Musiara Marsh. With growing excitement I read Jonathan's notes and realized that, in his account of the history of the Musiara pride, the idea of *The Marsh Lions* was staring us in the face.

The story is essentially a true one. All the central characters are real, and most of the incidents I have described actually happened, as Jonathan's photographs show. Wherever possible I have relied on Jonathan's personal observations, expanded by those occasions when I was fortunate enough to be able to watch the lions myself. Regular reports from drivers, camp managers and others who could recognize the individual animals helped to fill in the gaps.

Inevitably, there were times when no one was present, and it has been necessary to stray into the realms of fiction to maintain the narrative. The death of Dark Mane is a case in point – although his disappearance was a mystery, I used this instance as an example of the ignominious loss of a number of fine male lions before the ban on hunting was introduced by the Kenyan Government. At all times, however, I have tried to portray the animals and their world as faithfully as possible, guided by the wealth of scientific research which has done so much to increase our understanding of animal behaviour over the past twenty years.

To bring the sights and sounds and smells of wild Africa still closer I have tried to enter the skin of the main protagonists, to imagine how it feels to be the hunter – and the hunted. Of course, no one can ever truly see the world through the eyes of a lion, and if distortion or a shade of anthropomorphism has crept in, the fault is nobody's but my own.

Finally, if there is a message in this book, it is a plea to this generation, on behalf of the next, to continue to cherish the Mara as one of the most beautiful of the earth's wild places, and one of the last great strongholds of African plains game.

BRIAN JACKMAN,
Powerstock,
Dorset.

Photographer's Note

THE PHOTOGRAPHS in this book record in the majority of cases actual events described in the text. Animals in the Maasai Mara are generally extremely approachable. On occasion lions would seek the shade afforded by the vehicle I was in, or if this did not suffice would even take to crawling under the car. Young cheetahs are sometimes even bolder, jumping on to the roof, straddling open roof-hatches and biting windscreen wipers and tyres. I did not actively encourage this behaviour; it is simply a reflection of the lack of fear of vehicles that these animals have developed. It would be both foolish and untrue to infer that any of the animals were any the less wild or less entitled to respect. My only regret is that leopard do not share this lack of concern for vehicles. In general they are shy and very difficult to photograph, though fortunately for everybody Mara leopards now seem to be holding their own and certain individuals are beginning to tolerate relatively close observation.

All the pictures were taken using natural light. No hides were used to gain access to particular animals, the majority of photos being taken from a car, and a few from the air or on foot.

For photographing from a car, I built a metal table with two legs that slotted into brackets on the inside of my door when the window was lowered. This served as a tripod, and used in conjunction with bean-bags enabled me to rest my cameras and telephoto lenses in a firm and stable position.

Generally I preferred the lower angle of view afforded by shooting (photographing) from my door window. If necessary I could stand on the driver's seat and take pictures through my roof-hatch – this proved useful when long grass obscured the subject or if it was impossible to move the vehicle quickly to change angle. However, the elevated view through the roof-hatch tended to lessen the impact of the animal being photographed. In areas outside the reserve it is permitted to get out of your vehicle, which meant I could take pictures from ground level or under the car, which acted as a blind. This is especially effective for sunrises, sunsets and certain scenic shots; the low angle can also be very dramatic, as it allows the animal to dominate the photograph.

My equipment consisted of several Canon A1 cameras and motor drives, and the following lenses: 24mm f2.8; 35–70 mm f2.8 zoom; 85 mm f1.2; 135 mm f2; 200 mm f2.8; 80–200 mm f4 zoom; 85–300 mm f4.5 zoom; 400 mm f4.5.

I carried my equipment in a custom-built canvas seat-cover. This fitted over the double front seat adjacent to the driver's seat in my Toyota Landcruiser vehicle. The seat-cover consisted of four vertical compartments in each of which I placed a camera and lens. The four cameras were fitted with 35–70 mm zoom, 135mm, 200mm and 400mm lenses. I was thus able to change cameras and lenses quickly and also minimize the chances of missing a vital shot by running out of film in action situations. It also provided the added bonus of preventing cameras being dropped on the floor!

Dust can be a nightmare in Africa and I used chamois leathers, large brushes and dust-off cannisters to try and lessen its effects. A small particle of dust or grit can scratch film emulsion extremely easily, especially when using motor drives. Lens filters were used to protect lenses from dust and scratches. Film was stored in a cool-box or behind by seat where it remained relatively cool.

Where possible I used Kodachrome 64 and occasionally Kodachrome 25 slide film. These films are virtually grain-free and give high quality reproduction. There are, however, times when one is shooting in low light conditions or when high shutter speeds are needed to prevent blurred pictures in action sequences. There are basically two ways round these problems. One is to use high speed film such as Ektachrome 400 and 200, but they tend to be grainy and have a different colour balance to Kodachrome; the other is to buy fast lenses that can operate in these situations by using large maximum apertures. Canon have been the leaders in this field and produce a range of high quality compact optics that even allow one to photograph using telephoto lenses in these circumstances. Their latest lenses featuring these characteristics include an 85mm f1.2, a 300mm f2.8, and a 400mm f2.8 – the fastest lenses of their type in the world.

But I soon realized that the most important thing is to learn as much as possible about the habits and behaviour of the subject you are photographing. Only then does one start to anticipate where to position oneself and what lens is required *before* things start happening.

*   *   *

The drawings that precede each chapter were produced in pen and ink by a process known as stippling. This involved a preliminary pencil sketch to establish dimensions, after which the picture was built up by a series of ink dots produced by graphic drawing pens. A series of Rotring pens with nibs of varying dimensions (0.1, 0.2, 0.3) were used to produce a pattern of dots of different tones. The pencil marks were then erased with a soft rubber, leaving the ink intact.

To live and work in the Mara has been the greatest joy of my life – I only hope that a little of that is reflected in the photographs and drawings in this book.

Jonathan Scott
Maasai Mara
Kenya

Acknowledgements

Many people have been most generous with their time and assistance, providing hospitality and information in the course of the preparation of *The Marsh Lions*.

Brian Jackman would particularly like to thank:

John and Pat Eames, through whose generous hospitality he was able to visit Kenya so many times;

Julian and Jane McKeand, Don Hunt, Patrick and Patricia Orr, Esmond and Chrysee Bradley Martin, for their hospitality on numerous occasions;

George Adamson, who probably knows more about lions than anyone, and his right hand man, Tony Fitzjohn, for receiving him so kindly at *Kampi ya Simba* in the Kora Reserve, and for giving him his first experience of lions at close quarters;

Norman Carr, for imparting much valuable knowledge gained during a lifetime of living among wild animals in Zambia, and Cecil and Connie Evans, also of Zambia, for an exciting baptism in bush-bashing;

Hugo van Lawick and Alan Root, the finest African wildlife film-makers in the business, for their invaluable advice on the Serengeti and its wildlife;

Harold Evans, former editor of *The Times* and Frank Giles, editor of *The Sunday Times*, for allowing him time to finish the book, and his most heartfelt thanks to Richard Girling, also of *The Sunday Times*, for his unfailing encouragement and advice with the text.

David and Lida Burney for their invaluable thesis and help on the subject of cheetahs.

Jonathan Scott would especially like to thank:

Jack Block, who did so much to help and encourage him when he first arrived in Kenya, and made it possible for him to stay;

Tubby Block, who bought the first drawing that he sold, and thus encouraged him to produce more;

Joseph Rotich, who taught him so much about the Mara and its animals, and always seemed to find him, sooner or later, when he got stuck in the mud;

Roy Wallace, for untold kindness and assistance as manager of the Mara River Camp;

The Mitchell family for many kindnesses at Kiambethu Farm, Limuru;

Colonel T.S. Conner, who so generously made his house in Nairobi a second home, and who worked tirelessly to help him in every way;

Pippa Millard, whose help and hospitality in London were second to none;

Ann Olivecrona, who kept him informed of events in the Mara during his absence, and without whom the story would not have been complete;

Jock Dawson and Enid Phillips at Mara Buffalo Camp, who kept track of the hunting dogs and cheetah, quite apart from their cheerful welcome whenever he visited them;

Adrian and Vicky Luckhurst for their hospitality;

David Goodnow and Booth Henson for their generous assistance with equipment;

John Ole Naiguran, Senior Game Warden of the Maasai Mara and his staff;

The staff of the East African Herbarium for identifying plant material;

His family, especially his mother, who has helped and encouraged him in every way possible in doing what he enjoys most – working amongst wildlife.

Sadly Ian Ross and Annette Kampinga died before the completion of *The Marsh Lions*. They were not only good friends, but a much loved part of the Mara. Ian provided invaluable information on the characters involved in the story, and helped clarify the course of events in Jonathan's absence. Both their lives continue in spirit somewhere overlooking Paradise Plain, not far from the Mara River.

In addition, both authors would like to mention:

Jock Anderson of East African Wildlife Safaris, who gave Jonathan the opportunity to live in the Mara and is considered a good man to have around when being chased by an angry hippo;

Julian Larby and Alan Doig of Mara River Camp; Phil and Peggy Allen and Murray Levet of Governor's Camp and the management and staff of the Mara Serena Lodge, all of whom have made them welcome on a number of occasions and have helped to make their time in the Mara so very memorable;

Roger Houghton of Elm Tree Books, who must at one time have thought *The Marsh Lions* was the only book he was publishing, but who maintained an astonishing level of enthusiasm in the project and, with the help of Caroline Taggart, kept their morale boosted to the end;

Mike Shaw of Curtis Brown Ltd, their agent, a continual source of help and guidance, quite apart from his enthusiasm and encouragement;

Weidenfeld and Nicolson, for permission to quote the passage from Karen Blixen's *Letters from Africa*.

The very fact that it is still possible to view wildlife in the abundance that one finds in the Maasai Mara is a reflection of the determination of the Kenyan Government to preserve its wildlife heritage for future generations to enjoy. By pledging its active support to this effort, the rest of the world can play its part in ensuring a future for Africa's wildlife.